Sustainable Development Goals Series

The **Sustainable Development Goals Series** is Springer Nature's inaugural cross-imprint book series that addresses and supports the United Nations' seventeen Sustainable Development Goals. The series fosters comprehensive research focused on these global targets and endeavours to address some of society's greatest grand challenges. The SDGs are inherently multidisciplinary, and they bring people working across different fields together and working towards a common goal. In this spirit, the Sustainable Development Goals series is the first at Springer Nature to publish books under both the Springer and Palgrave Macmillan imprints, bringing the strengths of our imprints together.

The Sustainable Development Goals Series is organized into eighteen subseries: one subseries based around each of the seventeen respective Sustainable Development Goals, and an eighteenth subseries, "Connecting the Goals", which serves as a home for volumes addressing multiple goals or studying the SDGs as a whole. Each subseries is guided by an expert Subseries Advisor with years or decades of experience studying and addressing core components of their respective Goal.

The SDG Series has a remit as broad as the SDGs themselves, and contributions are welcome from scientists, academics, policymakers, and researchers working in fields related to any of the seventeen goals. If you are interested in contributing a monograph or curated volume to the series, please contact the Publishers: Zachary Romano [Springer; zachary.romano@springer.com] and Rachael Ballard [Palgrave Macmillan; rachael.ballard@palgrave.com].

Abraham R. Matamanda • Verna Nel
Editors

Sustainable Development Goals and Urban Health

Strides, Challenges and Way Forward for Poor Neighborhoods

Springer

Editors
Abraham R. Matamanda
Department of Urban
& Regional Planning
University of the Free State
Bloemfontein, South Africa

Verna Nel
Department of Urban
& Regional Planning
University of the Free State
Bloemfontein, South Africa

ISSN 2523-3084 ISSN 2523-3092 (electronic)
Sustainable Development Goals Series
ISBN 978-3-031-68733-4 ISBN 978-3-031-68734-1 (eBook)
https://doi.org/10.1007/978-3-031-68734-1

This Springer imprint is published by the registered company Springer Nature Switzerland AG
The registered company address is: Gewerbestrasse 11, 6330 Cham, Switzerland

Foreword

The sustainable development goals (SDGs) are at the centre of global development. Many cities are grappling to achieve even the basic set of goals and targets, yet in others, several milestones have already been achieved, especially in the Global North. This global disparity echoes one of the important themes of this book, inequities in the determinants of health. The challenges of achieving targets set in the SDGs mainly lie in the complexity of cities, which can make it difficult to identify effective measures to tackle specific goals. By their very nature, urban spaces influence population health outcomes in many ways, intersecting with the SDGs in multiple ways. For example, housing conditions, access to basic services, climate change and inequality are issues that highlight the interplay of the SDGs and urban health.

While it is true that many scholars have researched this nexus between urban health and the SDGs, this book situates that discussion within the South African context. This provides a critical lens for exploring deeper understandings. This book highlights concerns of a research gap of serious engagement and scholarship among urban scholars on the need to localise the debates about SDGs and urban health. More traditional nationally situated studies, although much needed, can lack the necessary granularity, resulting in certain local conditions, marginalised stakeholders or neighbourhood effects being overlooked. Based on these insights, the editors, Abraham Matamanda and Verna Nel, have compiled this collection, *Sustainable Development Goals and Urban Health: Strides, Challenges and Way Forward for Poor Neighborhoods*. Abraham and Verna are seasoned urban scholars whose scholarship focuses on urban governance, planning and management in the Global South, specifically the issues in southern African cities. They have applied their experience and knowledge in compiling this collection providing rich insights based on their previous work on medical geography in South Africa and Zimbabwe. Such work has tended to explore the issues surrounding pandemics and epidemics in cities, health inequalities, all resonating with the SDGs for the creation of safe, inclusive, sustainable and resilient cities and urban areas.

The collection of papers provides insights and an understanding of the local dynamics inherent in urban health within the context of the SDGs. The chapters highlight the urban health inequalities in South African cities and the institutional and structural factors contributing to these inequalities and the impact on the SDGs. Applying a healthy city and One Health approach, the chapters seek to uncover local dynamics inherent in the attempt to promote urban health, also showing that the SDGs cannot be considered linearly but need to be approached holistically. Through this approach, the papers in this collection help articulate issues and provide direction showing the reader where critical and substantive matters require more attention.

Using the case study of Mangaung Metro Municipality in South Africa, the edited volume significantly contributions to discussions on urban health and efforts to achieve sustainable development, using a healthy city approach. Several critical issues emerge from the book, making it a timely and valuable read for scholars, policymakers and the public.

Urban health is a significant challenge overwhelming Global South city, yet it remains overlooked by urban planners and geographers. South Africa, for example, has multiple urban health challenges including a high prevalence of HIV/AIDS, tuberculosis and many non-communicable diseases. While several studies have been undertaken focusing on the spatial aspects of some of these diseases, this volume brings a fresh perspective, situating this discussion in the context of the SDGs. Second, the authors adopt a local perspective in the analysis. This is critical in that it allows a "worm's view", articulating issues in a way that overcomes generalisation that often characterise national-level analysis or progress towards the SDGs.

Expert Advisor to the World Health Organization and UN-Habitat
Nairobi, Kenya
Editor-in-Chief, Cities & Health Journal
Bristol, UK

Marcus Grant, FFPH

Acknowledgements

Many individuals have made this book project a success. First, we appreciate the cooperation of the communities in the suburbs of Mangaung Metro Municipality who consented to participate in the interviews and provide insights into their lived experiences. These lived experiences enabled the authors to understand the urban health dynamics in MMM while reflecting on the sustainable development goals (SDGs).

Second, the grant from the Center for Graduate Support, Department of Research and Development at the University of the Free, which covered part of the costs of the fieldwork, is greatly appreciated. This funding enabled the fieldwork to be conducted which helped to generate the data for the respective chapters. The funding was also used to fund two workshops where the authors discussed and refined the respective chapters.

Third, the assistance with language editing from Liesl Van Der Westhuizen, the Chief Officer: Science Writer in the Office of the Dean, Faculty of Natural and Agricultural Sciences, is greatly appreciated.

Fourth, the following reviewers who reviewed the chapters in the book are greatly appreciated:

- DR. Takudzwa TARUZA, Department of Urban & Regional Planning, University of Kwa-Zulu Natal, Durban, Republic of South Africa
- Prof Walter MUSAKWA, Department of Geography, Environmental Management & Energy Studies, University of Johannesburg, Johannesburg, Republic of South Africa
- DR. Gabriel CAMARA, Department of Geography, "Alexandru Ioan Cuza" University of Iași, Iasi, Romania
- DR. Purity HAMUNAKWADI, Nelson Mandela Metropolitan University, Republic of South Africa
- DR. Joseph ZHAO, University of Glasgow, United Kingdom
- DR. Morgen ZIVHAVE, Department of Urban and Regional Planning, University of Botswana, Gaborone, Botswana
- DR. Durowoju S OLUFEMI, Department of Geography, University of the Free State, Bloemfontein, South Africa & Department of Geography

Contents

Contributors

Palesa Beloved Ayantokun is currently a master's degree student in Geography at the UFS, South Africa. She also holds an Honours and Bachelor degrees in Geography from the UFS. Palesa grew up in Bloemfontein also known as the "City of roses". Growing up in the city and observing all the changes that have occurred has been an influence on Palesa's research interests. She is passionate about research and her interests are on Urban Geography and Human Settlements. She has presented both her honours and master's research at academic conferences. Palesa is currently a part-time Geography Lecturer at the Central University of Technology, South Africa.

Johannes Bhanye is a Researcher and Academic in Migration and Urban Planning Studies in the Global South. He is currently a Post-Doctoral Research Fellow with the UFS, Department of Urban and Regional Planning. He also holds a B.Sc. (honours) degree in Rural and Urban Planning and an M.Sc. in Social Ecology, both from the University of Zimbabwe. His research interests cut across urban land governance, migration, human settlements, urban informality, cities and social change, climate change, governing the commons and rural development, among other development-related topics in Africa.

Natal Buthelezi is a registered architect with the South African Council for the Architectural Profession (SACAP) and holds a Ph.D. in Urban and Regional Planning from the UFS. His doctoral research, titled "Morphological characterisation of informal settlements and their adaptation to natural disasters in the face of climate change: A case study of Buffalo City, South Africa," demonstrates his devotion to urban resilience. Natal examines the complex dynamics of informal settlements and climate change adaptation with a focus on fostering sustainable development and resilience in vulnerable urban communities. His research work sheds light on potential transformative pathways, ultimately empowering vulnerable communities and propelling the advancement of resilient urbanisation by seamlessly integrating architectural expertise with urban planning principles.

Halleluah Chirisa holds a B.Sc. Hons Geography and Mathematics (Bindura University) and an M.Sc. in Population Studies (University of Zimbabwe). Currently Research and Innovation Director at Zimbabwe Ezekiel Guti University (ZEGU), Ph.D. student with the UFS in South Africa, Geography Department.

Innocent Chirisa the Acting Vice Chancellor of ZEGU is a Full Professor in Urban and Regional Planning. He has a keen interest in urban and peri-urban dynamics. Currently he focuses on environmental systems dynamics with respect to land-use, ecology, water and energy. He holds a D.Phil. in Social Sciences. Prof. Chirisa is also currently a Research Fellow with the Department of Urban and Regional Planning, UFS, South Africa. He has also served as a Visiting Professor at the Chinhoyi University of Technology (Zimbabwe). Membership of Professional Associations including: Fellow of the Pan-African Scientific Research Council (FPASRC); Member of the African Academy of Sciences (AAS); Member of the American Association of Geographers (AAG); Member of the American Political Science Association (APSA); Member of the Global Land Programme (GLP); Member of the Research and Development Committee of the Health Professions Committee (R&D-HPA (of Zimbabwe)); Member of the International Society for Urban Health (ISUH); Member of the International Society of City and Regional Planners (ISOCARP); Member of the Organisation for Social Science Research in Eastern & Southern Africa (OSSREA); Member of the Zimbabwe Institute for Urban and Regional Planners (ZIRUP).

Mischka Dunn is currently a Research Fellow for the University of the Free State, South Africa, and a Professional Teaching Fellow at the University of Auckland, New Zealand. In her previous role, she was an Urban Geography Lecturer at the University of the Free State. She holds an undergraduate degree in Geography and Environmental Management, an honours degree in Spatial Planning, a master's degree in Urban and Regional Planning and a Ph.D. in Urban and Regional Planning. Her research interests lie in alternative land use management systems for low-income areas, social and community wellbeing and child-responsive urban planning.

Tazviona Richman Gambe is a postdoctoral fellow at the Centre for Development Support, University of the Free State in South Africa. His research interests include urbanisation in Africa, regional economic resilience, firm innovation and resilience, urban water supply and spatial planning and development.

Mareli Hugo is a driven and successful Ph.D. candidate with a wealth of experience in urban and regional planning, GIS, research and academic fields. Having completed her honours in Spatial Planning and a master's degree in urban and Regional Planning, she has contributed to the academic field and public sector. As a Lecturer, Mareli is guiding and inspiring new minds within the field of spatial planning. Within the public sector, she has assisted in the development of Spatial Development Frameworks, Community Research and Relocation, Public Participation and Project Implementation. Currently, Mareli is serving as a Registered Candidate Town Planner (Registered with South African Council of Planners (SACPLAN)) and a part-time lecturer at the University of the Free State. Overall, Mareli is a driven academic, researcher, urban planner and GIS expert who has made significant contributions in a range of roles.

Simbarashe Jombo (Ph.D. Remote Sensing and GIS) is a geospatial data scientist currently serving as a postdoctoral research fellow in the Department of Physical and Earth Sciences at Sol Plaatje University. With more than 6 years of expertise, he excels in urban tree classification, land use and land cover mapping, bibliometric analysis, vegetation mapping and change detection. Proficient in various remote sensing and statistical software, Simbarashe seamlessly combines technology with environmental sustainability in his research. He has made significant contributions to studies on arid rangeland degradation, crop mapping, urban geography and changes in land use and land cover. His recent focus involves the use of machine learning and deep learning algorithms for classification to map and analyse maize production areas, with the aim of improving agricultural sustainability and food security.

Jennilee Kohima is a Spatial Planning Lecturer in the Department of Architecture, Planning and Construction at Namibia University of Science and Technology. She obtained her Bachelor of Technology in Land Management specialising in Land Use Planning from Polytechnic of Namibia, Master of Science in Land Management and Land Tenure from the Technical University of Munich and both Bachelor Honours in Spatial Planning and Masters in Urban and Regional Planning from UFS. She has over 15 years of work experience in the public, private, development and academic sectors in various development and planning fields. Currently, she is a Ph.D. candidate at UFS, and her research interests are on issues of urban informality, land use management, bottom-up planning, urban planning policy and practice, housing and governance, and planning education.

Lucia Leboto-Khetsi is a postdoctoral research fellow under the Centre for Development Support at the UFS, South Africa. She obtained her Bachelor of Arts in Urban and Regional Planning from the National University of Lesotho, her honours in Spatial Planning, and her master's and Ph.D. in Urban and Regional Planning from the UFS. She has over 16 years of experience within the non-governmental, private and academic sectors. She is affiliated with the SACPLAN, SAPI, International Geographical Union (IGU), ISOCARP and Regional Science Association International (RSAI). Her research interests include low-income housing, local economic development, local governance, urban health, urban planning, Africa, COVID-19, heritage geographies and climate change.

Timothy Lehobo is a Ph.D. candidate majoring in Urban and Regional Planning and a part-time lecturer at the University of the Free State. He is a Candidate Town and Regional Planner registered with the South African Council for Planners. His academic background has been in Urban and Regional Planning, from undergraduate studies through to the doctoral level. He has worked in housing, town planning, geographic information systems and academia. His research interests include housing, human (informal) settlements, township economies and spatial justice, among others.

Maureen Maisiri is a Ph.D. candidate at the Disaster Management Training and Education Centre for Africa (DiMTEC) at the UFS, South Africa. Her doctoral research focuses on food insecurity, disaster risk reduction and resilience in smallholder farmers in South Africa. She is particularly interested in exploring how building resilient food systems aids in meeting the United Nations Sustainable Development Goals (SDGs) as well as address the priorities of the Sendai Framework of Disaster Risk Reduction. She holds a master's degree in Post-Harvest Science and Technology from Chinhoyi University of Technology, Zimbabwe, and a bachelor's degree in Food Science and Nutrition from Midlands State University, Zimbabwe.

Fortune Mangara is a highly qualified researcher with extensive spatial and urban planning expertise. He holds a degree in Rural and Urban Planning from the University of Zimbabwe and an M.Sc. and Ph.D. in Urban and Regional Planning from the North-West University. Dr. Mangara is a post-doctoral research fellow at the School of Ecological and Human Sustainability at the University of South Africa, Department of Geography. His research interests encompass a wide range of topics, including spatial and urban resilience, disaster risk studies, disaster resilience, disaster risk governance, urban disaster risk, social and economic geography, urban policy, urban governance, transport planning and mobility and spatial transformations in African cities.

Tafadzwa Clementine Maramura She holds a Ph.D. in Public Management and Water Governance from the North-West University (*Valedictorian*), a Master of Public Policy (*Valedictorian*), Honours Public Policy and Administration (*Cum Laude*) and a B. Soc. Sci (*Cum Laude*) from the University of Fort Hare. She is currently a Senior Lecturer at the University of the Free State. Her research interests are in Water Governance, Sustainable Service Delivery, Sustainability and Public Policies spanning across Africa. She is a Brightest Young Minds (BYM) in Africa alumnus, where she is working towards the promotion of research transparency and reproducibility in Africa since 2017.

Abraham R. Matamanda (Ph.D. Urban & Regional Planning) is an NRF Y2-rated Urban and Regional Planner. Abraham lectures at the University of Free State (UFS) and is also affiliated with the SARChI Chair on City Region Economies in the faculty of Economic and Management Sciences at UFS. Currently, he serves as the editor of the *Town Planning Journal* published by UFS and serves on the academic editorial board of *Plos Water Journal*. Abraham is also the current Chairperson of the Free State Chapter of the South African Planning Institute (SAPI). He is a fellow of the Department of Higher Education and Training (DHET) Future Professorate Programme Phase 1, third Cohort. Abraham is the South African PI for a global collaborative research project exploring how children and young people from monetary poor households have adapted to the COVID-19 pandemic focusing on food, education and play/leisure (https://panexyouth.com/). His research focuses on urban governance and planning, climate change adaptation and resilience, informal Global South urbanism, urban land markets and housing studies.

Leemisa Simon Matooane is an accomplished Ph.D. candidate exploring the nexus between Food Systems and Urban Planning and Management. He has a great deal of expertise in urban and regional planning, GIS and programme administration. Leemisa's experience and talents were further acknowledged when he was appointed to Habitat for Humanity Africa's steering committee, where he can put his abilities to use supporting the vital work that the organisation does throughout the region. Leemisa is also having a big impact on academia as a part-time lecturer in the University of the Free State's Department of Urban and Regional Planning and as a publishing researcher. In addition, he has published his work that he has presented at international conferences. He is an extremely talented, accomplished and motivated urban planner, GIS specialist and programme manager who has made significant contributions in a range of roles and continues to do so.

Abongile Mgwele is currently a Lecturer, Ph.D. candidate and emerging scholar with the Department of Urban and Regional Planning at the UFS. She has a Spatial Planning Honours and a master's in Urban and Regional Planning. She is also affiliated with the South African New Generation of Academics Programme (nGAP) and the SACPLAN. Her research interests cut across spatial governance, power and informality in Africa; unlocking the potentiality of rural planning; gender mainstreaming; and sustainability in urban planning and marginalised housing practices and socio-cultural aspects. She advocates putting the marginalised in the centre of urban and regional planning.

Kgosi S. Mocwagwae (Ph.D. Urban & Regional Planning) is an emerging researcher in the field of Urban and Regional Planning. Kgosi is a Professional Planner, registered with SACPLAN. He currently lectures at the UFS, where he is also the current Academic Head of the Department. Further to his work at the UFS, Kgosi currently serves in two municipal planning tribunals as a Deputy Chairperson and member and is the Deputy Chairperson of the Free State Chapter of SAPI. In 2023, Kgosi was awarded the overall first place and winner of the Academia at the 2023 Young Planner Awards by SACPLAN. Kgosi's current areas of focus include planning for basic service provision, rural and traditional planning, and sociocultural aspects in planning.

Thulisile Ncamsile Mphambukeli is an NRF-rated international scholar and currently a Head of Department and Associate Professor in the Department of Urban and Regional Planning, University of Johannesburg. She is a co-editor of the *Journal of BRICS Studies*, and a Review and Topic Editor for Governance and Cities for the *Frontiers in Sustainable Cities* journal. She is a member of the South African BRICS Think Tank. Her research interests include social justice in planning, climate change and agriculture; coloniality and informality; human security and humanitarian response; integrated transport planning; spatialisation of blackness; and urban anthropology. She recently published on urban insecurity and social justice in post-colonies, financing regenerative agricultural practices, sustainable development, planning for sustainable quality infrastructure, applied systems analysis and apartheid.

Charles Poleni Mukumba is a postdoctoral research fellow at the Centre for Development Support, University of the Free State. He completed his Ph.D. in Construction Management, investigating Public Construction Procurement and SME Development from the University of the Free State. Charles studied Architecture at the Copperbelt University, Zambia. His Master of Land and Property Development Management with specialisation in Housing at the University of the Free State focused on participatory approaches to upgrading informal settlements. His current research highlights urbanisation trends, housing and informal settlements in mining communities on Zambia's Copperbelt.

Verna Nel qualified as a town and regional planner at Wits University and obtained her M.Sc. and Ph.D. through UNISA. After three decades of working primarily in municipalities, she moved to the Urban and Regional Planning Department of the UFS. She is also affiliated with the SARChI Chair on City Region Economies in the Faculty of Economic and Management Sciences at UFS. She is an NRF C2-rated scholar. Besides postgraduate supervision, Verna has written one book and co-edited four books which reflect her broad interests in planning. These include spatial governance, secondary and mining cities, local economic development and spatial resilience.

Moleboheng B. Pherane is a Geography Ph.D. candidate at the University of the Free State. Moleboheng lectures part-time at the University of the Free State in the Department of Geography. She majored in Geography and Sociology in her junior degree, and therefore, her research focuses on blending the two disciplines. Her research focuses on urban ecology and how it is influenced by different societies. More specifically, she focuses on wetland degradation, education, sustainability, environmental humanity and ecological management.

Tebello Putsoane is a Ph.D. candidate in the Department of Urban and Regional Planning at UFS. She researches on the interplay of climate change and ecologically sensitive areas in Lesotho.

Siphokazi Rammile is a Lecturer at the Department of Urban and Regional Planning at the UFS. She is a full-time academic with a substantial background in private practice. As a Ph.D. student in Urban and Regional Planning, the field from which she obtained her master's degree, her research interests are routed in urban dynamics. The influence of her undergraduate degree in geography and environmental management remains evident through her interest in toponomy, sustainable development and nature-based solutions. Her current research interests include liveability, tactical urbanism, urban spatial displacement and the relationship between well-being and urban planning.

Adriaan Van Der Walt is a Senior Lecturer in Physical Geography and Geographic Information Systems in the Department of Geography at the UFS. In 2020, he obtained his Ph.D. in Geography from the School of Geography, Archaeology and Environmental Sciences at the University of the Witwatersrand. His primary research field is situated in biometeorology, exploring the impacts of climate change on both the natural environment and human populations. Dr. van der Walt co-authored a publication in the *South African Journal of Science,* titled "Statistical classification of South African Seasonal Divisions on the Basis of Daily Temperature Data", which garnered widespread media coverage with online articles in *Business Insider Travel* and *UFS News*, local and national newspaper articles in *Rapport*, *The Citizen* and *Cape Argus*, radio interviews with *SAFM*, *LotusFM* and *RSG Monitor*, an article in *Landbouweekblad magazine* and a podcast interview with *The Conversation Africa* on extreme temperatures in South Africa. Most recently, Dr. van der Walt was interviewed by *KykNET Verslag* en *Fokus* on the flooding events that took place in KwaZulu-Natal, *Sunday Times* and *Netwerk24* on the possibility and effects of increasing heat wave events in southern and South Africa and published a co-authored article in *QUEST: Science for South Africa* on research in Africa to tackle the climate change crises. His research has been recognised by the Society of South African Geographers' Centennial Award for Emerging Career Researchers (2020–2021). Recently, he was invited to present an online Master Class at the 8th South Asia Heat Health Information Network (SAHHIN) on *Heat Stress Mitigation and Adaption Perspectives in South Africa.* Furthermore, Dr. van der Walt is an elected executive council member of the Society of South African Geographers (SSAG), the SSAG Student and Young Professionals (SnP) group and the Global Climate Emerging Research Network (GCERN). He has been appointed as a member of the Editorial Board of the *International Journal of Biometeorology*, published by Springer, and was selected to be part of the Associate Editor Mentee Programme of the *South African Journal of Science*.

About the Editors

Abraham R. Matamanda (Ph.D. Urban & Regional Planning) is an NRF Y2-rated Urban and Regional Planner. Abraham lectures at the University of Free State (UFS) and is also affiliated with the SARChI Chair on City Region Economies in the faculty of Economic and Management Sciences at UFS. Currently, he serves as the editor of the *Town Planning Journal* published by UFS and serves on the academic editorial board of *Plos Water Journal*. Abraham is also the current Chairperson of the Free State Chapter of the South African Planning Institute (SAPI). He is a fellow of the Department of Higher Education and Training (DHET) Future Professorate Programme Phase 1, third Cohort. Abraham is the South African PI for a global collaborative research project exploring how children and young people from monetary poor households have adapted to the COVID-19 pandemic focusing on food, education and play/leisure (https://panexyouth.com/). His research focuses on urban governance and planning, climate change adaptation and resilience, informal Global South urbanism, urban land markets and housing studies.

Verna Nel qualified as a town and regional planner at Wits University and obtained her M.Sc. and Ph.D. through UNISA. After three decades of working primarily in municipalities, she moved to the Urban and Regional Planning Department of the UFS. She is also affiliated with the SARChI Chair on City Region Economies in the Faculty of Economic and Management Sciences at UFS. She is an NRF C2-rated scholar. Besides postgraduate supervision, Verna has written one book and co-edited four books which reflect her broad interests in planning. These include spatial governance, secondary and mining cities, local economic development and spatial resilience.

Acronyms

AAG	American Association of Geographers
AAS	African Academy of Sciences
AI	Artificial Intelligence
AMR	Annual Mean Rainfall
AMT	Annual Mean Temperature
ANC	African National Congress
APSA	American Political Science Association
ASTM	American Society for Testing and Materials
B-BBEE	Broad-Based Black Economic Empowerment
BDRR	Blue Drop Risk Rating
BYM	Brightest Young Minds
CAP	Common Alerting Protocol
CHRIPS	Climate Hazards Group IntraRed Precipitation with station rainfall data
CoGTA	Cooperate Governance and Traditional Affairs
DHET	Department of Higher Education and Training
DIMTEC	Disaster Management Training and Education Centre for Africa
DIY	Do It Yourself
ET-SCI	Expert Team on Sector-specific Climate Indices
GCERN	Global Climate Emerging Research Network
GDP	Gross Domestic Product
GEE	Google Earth Engine
GHS	General Household Survey
GLP	Global Land Programme
FAO	Food and Agriculture Organisation
HIA	Health Impact Assessments
HiAP	Health in All Policies
ICT	Information and Communications Technology
IDP	Integrated Development Plan
IGU	International Geographical Union
IoT	Internet of Things
IPCC	Intergovernmental Panel on Climate Change
ISOCARP	International Society of City and Regional Planners
LULC	Land Use and Land Cover
MDG	Millennium Development Goals
Metro	Metropolitan Municipality

MMM	Mangaung Metropolitan Municipality
MSDF	Mangaung Service Delivery Forum
NBN	National Broadband Network
NCCRP	National Climate Change Response Policy
NDP	National Development Plan
NEHAWU	National Education Health and Allied Workers Union
nGAP	New Generation of Academics Programme
NGO	Non-governmental Organisation
NOAA	National Oceanic and Atmospheric Administration
NRF	National Research Foundation
OECD	Organisation for Economic Co-operation and Development
OSSREA	Organisation for Social Science Research in Eastern & Southern Africa
FPASRC	Pan-African Scientific Research Council
RDP	Reconstruction and Development Programme
RSAI	Regional Science Association International
SACAP	South African Council for the Architectural Profession
SAHHIN	South Asia Heat Health Information Network
SAICE	South African Institution of Civil Engineering
SANS241	South African National Standard 241
SAPI	South African Planning Institute
SDBIP	Service Delivery and Budget Implementation Plan
SDG	Sustainable Development Goal
SPLUMA	Spatial Planning and Land Use Management Act
SSA	Sub-Saharan Africa
SSAG	Society of South African Geographers
UFS	University of Free State
UN DESA	United Nations Department of Economic and Social Affairs
UN	United Nations
UNDP	United Nations Development Programme
UNSDCF	United Nations Sustainable Development Co-operation Framework
VIP	Ventilated Improved Pit toilets
WASH	Water, Sanitation and Hygiene
WHO	World Health Organisation
WMO	World Meteorological Organisation
WOS	Web of Science
$WSDI_n$	Warm Spell Index
ZEGU	Zimbabwe Ezekiel Guti University
ZIRUP	Zimbabwe Institute for Urban and Regional Planners

1 Introduction

Abraham R. Matamanda and Verna Nel

1.1 Introduction

Cities have always grappled with health problems (Mumford 1961; Barton 2009; Corburn 2013). The emergence and persistence of urban health problems are attributed to natural and Anthropocene factors. Most of the urban health problems have been linked to the rapid urbanization that has resulted in the increased concentration of people in cities (Corburn 2009; Kuddus et al. 2020; Pinchoff et al. 2020). With the concentration and densification of people in cities, there have been some unsustainable practices. For example, the disposal and management of solid waste and faecal matter in cities have led to cholera and typhoid outbreaks. An example has been the cholera outbreak that plagued London in 1854, resulting from communities using contaminated water (Bingham et al. 2004; Caplan et al. 2020). Albeit advances in portable water supply in cities, cholera pandemics persist in many African countries. Tulchinsky (2018) pointed out that waterborne disease is given less recognition as water and sewerage treatment standards are assumed to be part of modern living. However, there have been persistent cholera and typhoid outbreaks in Zimbabwe and Zambia due to limited water resources in urban areas (Mengel et al. 2014; Chigudu 2020; Africa CDC 2024). These diseases occur when governments fail to maintain the water and sanitation systems. In Zimbabwe, the situation was dire in 2008/09 when the mortality for the cholera pandemic was more than 4000 people, primarily concentrated in the country's large cities (Gwenzi and Sanganyado 2019). The case of Ebola in East and West Africa also depicts the public health pandemic that ravaged the region and was made worse by the porous borders, which brings in the perspective of 'global urbanisation' as articulated by Haase et al. (2018).

Many authors have also attributed the density of urban areas to spurring the emergence and contagion of certain infectious diseases (Jackson 2003; Vlahov et al. 2007). The compactness of cities where people are in constant interaction makes cities contagious spaces for urbanites—creating risks of infection when using public transport, visiting markets, sporting events and so on (Ahsan 2020). Especially at risk are citizens who reside in marginal spaces such as informal settlements. They are detached from formal services such as water and reticulated sewer, which compromises their living environments and ultimately, their health (Matamanda et al. 2024). In these instances, the urban living environment directly impacts urban health through the exposure of citizens to residing in overcrowded and

A. R. Matamanda (✉) · V. Nel
Department of Urban & Regional Planning, University of the Free State, Bloemfontein, South Africa

Center for Development Support, University of the Free State, Bloemfontein, South Africa

A. R. Matamanda, V. Nel (eds.), *Sustainable Development Goals and Urban Health*, Sustainable Development Goals Series, https://doi.org/10.1007/978-3-031-68734-1_1

Fig. 1.1 Solid waste accumulating in stagnant water creating a breeding ground for disease-causing agents in Botshabelo, South Africa. (Source: Authors 2023)

ill-ventilated housing disease-causing agents. This has been characteristic of most informal settlements and townships where the poor reside, where the physical environment contributes to outbreaks of diseases such as cholera, typhoid and malaria (Mathaai 2009). Disease outbreaks have occurred in some South African cities where residents in informal settlements lack access to water, sanitation and waste management (see Fig. 1.1). This situation aligns with SDGs 1 focusing on ending poverty, 6 relating to water and sanitation access and 11 on creating safe, sustainable, resilient and inclusive human settlements and how this impacts urban health.

Housing shortages in many African cities occur due to populations increasing so fast that neither safe housing nor job creation can keep up, resulting in overcrowding as the poor seek cheaper housing alternatives, which exposes them to respiratory and other diseases (Matamanda and Nel 2021). The quality of materials used to build some housing in some poor neighbourhoods exposes them to harsh weather. As shown in Fig. 1.2 corrugated iron sheets are used to construct housing in most informal settlements, which exposes the occupants to extreme heat in summers and the freezing winters experienced in the Free State Province in South Africa. Furthermore, the impacts of overcrowding and the squalid living conditions among the poor on urban health were visible during the COVID-19 pandemic, where the poor have suffered much (Wilkinson 2020; Matamanda et al. 2022a).

Urban lifestyles have affected people's diets, entertainment, and cultural activities, which have a bearing on their health. This critical issue relates to the food system and the availability of nutritious food (Dixon et al. 2007). The places where people obtain food have a bearing on urban health since some food is sold in open markets that may contaminate or sell such food. The handling of the food can result in its contamination (Battersby and McLachlan 2013; Tacoli 2020; see Fig. 1.3). The desperately poor may be forced to 'scavenge' for highly risky food, while some households may skip meals thus contributing to malnutrition and stunting, especially for children (Kimani-Murage et al. 2014). In South Africa, approximately one in ten people go hungry every day, hence malnutrition levels tend to be high (Manning 2021). Consequently, the hunger problem in South Africa due to malnutrition levels that are associated with stunting and obesity contributes to a major health crisis (Labuschagne 2022).

Some emerging scholarship on urban health in African cities focuses on the impacts of climate change (Kjellstrom 2016), a complex and growing problem that many planners are still grappling with. Heat waves, flash floods, and various other climate extremes have been a major challenge that has affected many communities and demand adaptation and mitigation strategies among the communities affected by these climatic effects (Patz et al. 2014). Planners and other development professions must pay attention to climate change impacts and their associated health effects, including heat stress, an increase in respiratory and cardiovascular diseases, injuries, and premature deaths (Ramin and McMichael 2009; Brousse et al. 2019). According to the World Meteorological Organization (WMO) (2023), between 1970 and 2021, extreme climate and water-related events caused 11,778 reported disasters, with just over two million deaths (https://twitter.com/WMO/sta-

Fig. 1.2 Makeshift materials used to construct shack in Botshabelo, South Africa. (Source: Authors 2023)

Fig. 1.3 Food waste collecting at the Jambanja fresh fruit and vegetables market in Chitungwiza, Harare. (Source: Authors 2022)

tus/1660551143829667842). African cities are at great risk as evidenced by the increasing infectious diseases. Malaria now occurs in certain parts that were previously not prone to mosquito-borne diseases as shown by the study by Mabaso et al. (2006). Incidences of diarrheal disease have been increasing in southern Africa due to increases in average temperatures and rainfall and children under five years have been at most risk (Viglione 2023). Heat-related illnesses have been reported to the on the increase owing to more frequent extreme heat events across South Africa in cities such as Bloemfontein, Johannesburg, Durban and Cape Town (Jones et al. 2022).

These urban health problems inherent in African cities have been attributed to factors such as rapid urbanization, climate extremes, urban penalty of the poor, misgovernance and corruption, collapse of the public health sector and other natural events such as floods and droughts. Several strategies have been employed (with varying levels of success) to mitigate these urban health problems and enhance the resilience of the cities to health pandemics. The rationale has always been to create habitable urban spaces where citizens have access to a good standard of living resonating with the sustainable development goals (SDGs) 3 and 11 aspiring and envisaging for the creation of healthy spaces that support human well-being.

1.2 Healthy Cities

The concept of healthy cities is critical; hence the need for this book with the increasing concentration of Africans in cities. With increasing urbanization rates in Africa, the impact on urban health is greater, with more issues that compromise living conditions, thereby making cities health hazards for the poor. As explained in the earlier section, the incidence of disease in cities has surged, hence the need to drive for the establishment and management of healthy cities.

According to the World Health Organisation (WHO) a healthy city "*is one that continually creates and improves its physical and social environments and expands the community resources that enable people to mutually support each other in performing all the functions of life and developing to their maximum potential*" (Hancock and Duhl 1988: 41). Consequently, developing healthy city is a complex issue that is difficult to achieve mainly due to the complexity of cities and the meshing of urban living and health issues. In attempts to explain the dimensions of the healthy city, the WHO has argued that it is best to conceive the healthy city as a process and not an outcome. In this regard, achieving a healthy city becomes an ongoing process with continuous evaluations meant to enhance the well-being of citizens while improving the standards of living. Eleven characteristics of a healthy city have been identified by WHO as tabulated in Table 1.1.

Table 1.1 WHO's 11 characteristics of a healthy city

Characteristic	Remark
1. Clean, safe, high-quality physical environment (including housing quality)	This characteristic resonates with the efforts of the Health in Towns Commission that was established by the British government in 1843 to bring sanity, order and improved living conditions in cities. Many cities now strive to improve the physical environment within cities.
2. A stable ecosystem now and sustainable in the long term	The natural environment has been identified to contribute to psychological and mental well-being hence critical for healthy cities.
3. Strong, mutually supportive and non-exploitative community	Community engagement and support of healthy city initiatives is critical as this increases the buy-in to the programmes.
4. A high degree of public participation in and control over the decisions affecting one's life, health and well-being	Citizen participation ensures that citizens contribute to matters that affect their well-being. Full participation is encouraged as the best form of participation.
5. Meeting of basic needs (food, water, shelter, income, safety, work) for all the city's people.	Accessing the basic needs helps to alleviate individuals and households from poverty and also diseases of poverty.
6. Access to a wide variety of experiences and resources with the possibility of multiple contacts, interactions and communication	Sharing experiences, opportunities and resources is a significant contributor to healthy cities as it fosters mutual understanding among individuals and groups in cities.
7. A diverse, vital and innovative city economy	A robust economy empowers citizens through different ways that relate to public health including the ability to pay for health services and insurance.
8. Encouragement of connectedness with the past, with the cultural and biological heritage and with other groups and individuals.	Situating the city within the broader historical context allows for a deeper and more nuanced understanding of the city dynamics.

(continued)

1.3 Why the Sustainable Development Goals Lens for This Book

The preceding sections confirm that urban health is a multifaceted phenomenon transcending urban socio-economic and political spheres. Based on the eleven characteristics of a healthy city articulated by WHO, it emerges that cities are complex spaces where urban health manifests in different aspects of the city and at different scales. Given the wide range of issues that determine urban heath, the broad set of goals set by the UN in the SDGs have been determine to be the most appropriate lens.

Table 1.1 (continued)

Characteristic	Remark
9. A city form that is compatible with and enhances the above parameters and behaviours	Different urban forms correlate with disease spreading, access to health care facilities and social well-being of the citizens.
10. An optimum level of appropriate public health and sick care services accessible to all	The availability of health infrastructure within specific thresholds is a great enabler of public health
11. High health status (both high positive health status and low disease status).	A healthy citizen contributes to a healthy and efficient city

Fig. 1.4 Summary of the 17 SDGs

In 2000, world leaders met to discuss ways to combat global poverty in its different dimensions. This meeting led to a broad vision that was translated into eight goals—Millennium Development Goals (MDGs). The MDGs became an overarching development framework for the world between 2000 and 2015 when these goals were superseded by the sustainable development goals (SDGs) in 2016 at a meeting the United Nations. These goals are a transformative agenda to guide global development with targets set for 2030. The SDGs consist of seventeen goals each with its own targets aimed at guiding, monitoring and advancing global development (see Fig. 1.4).

Despite the commitment and efforts by governments to achieve the SDGs by 2030 and 'not leave' anyone behind, poverty and inequalities remain entrenched in many countries (Ford 2023). Therefore, red flags have been raised on several SDGs that seem too ambitious and challenging to achieve within one and a half decades (Kroll et al. 2019; Leal Filho et al. 2020). Health challenges are among the persisting and growing challenges plaguing communities, especially cities in sub-Saharan Africa. Such health challenges are most acute among the extremely vulnerable groups who are visibly left behind in all aspects of urban health due to their marginalization.

The direct implications of cities on health are best revealed through the infrastructure and services that support health care. The presence of this health infrastructure is a direct measure of quality health care or the lack of it (Matamanda and Nel 2021). This perspective brings into account the need for the equitable distribution and provision of primary healthcare infrastructure and services to communities. However, the circumstances in most poor neighbourhoods in

African cities is the absence of clinics, hospitals or health emergency centers, a situation that compromises the health and well-being of citizens. In a report by Amnesty International on Hopley Farm Settlement in Harare, it was revealed that maternal mortality was increased due to the absence of primary health care facilities in the area which negatively affected pregnant women. In South Africa, the same situation is experienced where the poor in urban areas are not adequately serviced with primary healthcare facilities (Harris et al. 2011; Matamanda et al. 2022b). Access to healthcare infrastructure and services thus resonates with SDG 3 on health and well-being, SDG #9 on industry, innovation and infrastructure and SDG 11 on human settlements and their sustainability, inclusivity and resilience. Physical infrastructure, particularly clean water and decent sanitation is also a critical determinant of urban health.

The focus on locale is purposive and has been inspired by the works of Campbell and Cowan (2016) on radical incrementalism. Although the SDGs are a global commitment by world leaders, their success is transformative when the local issues and context are considered. This permits identifying the specific challenges plaguing local communities, the adaptive strategies and incremental steps and activities they employ to enhance their well-being. Understanding these small steps helps to map what works and what does not work for communities, thus providing a launch-pad for the formulation of policy initiatives centred on the communities' needs and aspirations.

1.4 Aim and Objectives of the Book

The main aim of this book is to understand the interplay between the SDGs and urban health. This is a critical issue in cities, considering the complexity of health issues and how they transcend across most SDGs. However, the SDGs are premised on a broad set of indicators and targets that are generalized. At the same time, the local contexts are always different and hence challenging to have a one-size-fits-all understanding of urban health problems. The nature of the SDGs thus calls for Health in All Policies (HiAP)—"an approach to public policy across sectors that systematically takes into account the health implications of decisions, seeks synergies and avoids harmful health impacts to improve population health and health equity". Therefore, adopting the HiAP, the book adopts a case study approach and considers the poor neighbourhoods of South African secondary cities, specifically, Bloemfontein, a metropolitan city. Each chapter presented in the book is premised on a particular SDG and how it relates to urban health. The following SDGs are analyzed in light of urban health: SDG 1 no poverty, SDG 2 no hunger, SDG 3 health and well-being, SDG 4 quality education, SDG 5 gender equality, SDG 6 clean water and sanitation, SDG 13 climate action, SDG 14 & 15 protecting the environment to reduce destruction and SDG 17 on peace, justice and institutions. Specifically, the objectives of the book are:

- To interrogate the specific targets of the respective SDGs relating to urban health and analyzing the SDG Index Score at the national and local level,
- To examine the set targets on the respective SDGs and the progress made thus far,
- To document the lived experiences and realities of the specific communities regarding the respective SDGs and urban health situation, and
- Explore the nexus of the respective SDGs to urban health, paying attention to enabling and disabling factors to achieve the envisaged targets.

1.5 Contextualizing the Mangaung Metropolitan City-Region

Mangaung Metro Municipality (MMM) lies in the Free State province, one of the nine provinces of South Africa. It is centrally located within and is the provincial capital of Free State Province.

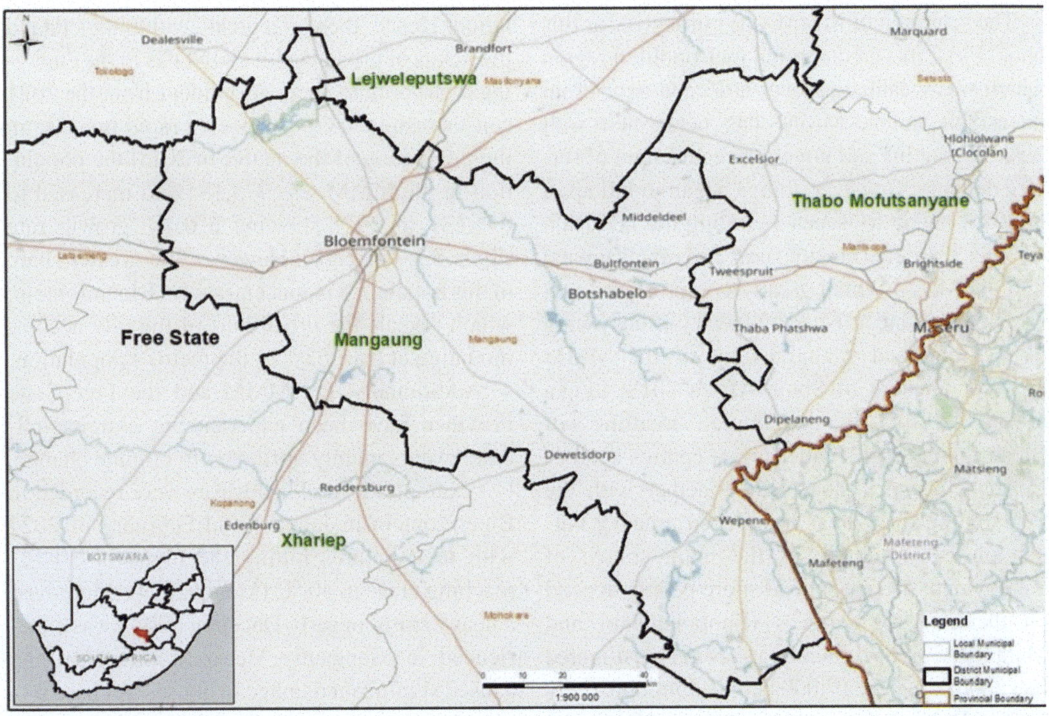

Fig. 1.5 Map of Mangaung Metro Municipality and the surrounding districts

As shown on Fig. 1.5, MMM also shares its south-eastern borders with Lesotho. The MMM covers an area of 9886km^2, which is mostly agricultural land. MMM is a Category A Municipality, which according to section 155(1)(a) of the South African constitution is "*a municipality that has exclusive municipal executive and legislative authority in its area*" (Republic of South Africa 1996). Mangaung Municipality was initially formed in 2001 when Bloemfontein the Judicial Capital of South Africa and the apartheid settlements of Botshabelo and Thaba Nchu were merged to form the new municipality. In 2011 Mangaung was declared a metropolitan municipality. There are four additional administrative regions in the form of small towns in the metro that include Dewetsdorp, Wepener, Soutpan and Van Stadensrus. This makes it a metropolitan region that is still in transition and grappling with multiple challenges including governance, service delivery and budgetary issues.

Since its inception, the metro has been marred by several maladministration issues that greatly compromised service delivery and development. Corruption and embezzlement of public funds in the municipality was so extensive that the state placed the MMM under administration. This book thus explores how such maladministration has derailed the progress towards the SDGs within the metro and the subsequent implications on urban health.

The MMM is characterized by a relatively well-developed economy largely based on public administration and defence activities (30.64% of total industrial share), 12.32% dominated by wholesale and retail trade and with education coming third with a share of 9.32%. These statistics obtained from the spatial tax data of 2022 confirms the heavy reliance of the Mangaung economy on service industries. Moreover, there are marked inequalities in the distribution of the economic activities which is partly explained by the legacies of the Apartheid planning which concentrated all economic activities in urban centres which in this instance benefited Bloemfontein making it the economic hub of the local and regional economy.

The selection of MMM was purposive for this book. First, it is an emerging metropolitan region which was established over a decade ago. From its establishment, MMM has performed well regarding audits and governance and state of service delivery. However, things began to fall apart in 2016, owing to issues including the inclusion of other districts (mostly rural and small towns) into the MMM which greatly compromised service delivery and affected budgetary issues owing to the increased spatial extent of the MMM (Auditor General of South Africa 2018; Lenka et al. 2021). For example, merging of Bloemfontein and smaller urban centres such as Botshabelo and Thaba Nchu together with the surrounding rural areas has placed a heavy burden on Bloemfontein as the main source of employment and revenue as more people depend on the city for services (Subramanyam and Marais 2022). Specifically, it has been estimated that approximately 17,000 people commute daily from Botshabelo and Thaba Nchu to Bloemfontein. Subsequently, there have been evidence of poor service delivery, misgovernance and impact on urban health and any strides towards meeting the SDGs.

Second, MMM was placed under provincial administration in 2019 and eventually under natial administration following a series of protests, misgovernance and of and poor service delivery which required the national intervention. This situation is riddled with multiple challenges that impact on the performance and governance issues of the metro hence the need to position the urban health and SDGs dynamics in this Metropolitan Municipality.

Furthermore, there are few studies focusing on emerging metros and the SDGs in South Africa. Marais and Twala (2021) have indicated that especially ignored and overlooked in the uran literature in South Africa is Bloemfontein the provincial capital of the Free State and major city in the MMM. Most studies focus on larger metropolitan municipalities (metros) such as Cape Town and those in Gauteng hence there is limited empirical understanding of the dynamics occurring in the smaller metros. Yet they are projected to absorb a larger share of the urban population. Hence, there is a great need to understand the nexus of urban health and SDGs in the emerging metropolitan areas. As evident from the 2021 census results, there has been a rapid increase in the population of the metro. In 2011, the population in the MMM was 775,184 and increased to 811,431 in 2021 showing a 0.4% growth rate (Stats SA 2022: 20). However, the greater share of this population is concentrated in Bloemfontein, which reveals the urban bias of the city accommodating at least 63% of the metro's population.

Additionally, the MMM and the Free State Province have been experiencing extreme climate events mainly attributed to climate change. For example, several heatwaves were recorded in Bloemfontein in January and February of 2024 with temperatures hitting the highs of 30 °C, reaching close to 40 °C (Mentz 2024). Moreover, a heavy storm in early December 2023 was experienced in Mangaung Metro Municipality and caused significant damage to houses in Phase 6 (a marginal settlement in Bloemfontein) and that killed a seven-year-old boy while leaving many residents injured (Semela 2023).

1.6 Outline of the Book

The book is structured into 11 chapters. This chapter introduces the study by articulating the urban health problem and how it compromises the attainment of healthy cities. Healthy cities are envisaged as they promote the well-being of citizens and enhance the liveability of cities. This discussion is situated in the context of the SDGs set by world leaders as global indicators and measures of progress and development across nations. Several targets have been set that relate to urban health, and this book focuses on this nexus by examining selected SDGs. The selection of South African case study has been purposive and helps understand the enablers and disablers of attaining healthy cities and the SDGs through local perspectives and experiences.

In Chap. 2 Natal Buthelezi, Lucia Khetsi-Leboto and Verna Nel situate the urban health problems in the context of SDG 1 which envisages ending poverty by 2030. They provide working definitions of poverty, its measure in the South African context and the strides to alleviate this social ill. The health dimension in this SDG focusing on poverty is articulated by analyzing the residents' inability to access basic services critical to sustaining a decent life and maintaining a healthy lifestyle. They posit that due to poverty, the poor are forced to engage in unhealthy practices that are life-threatening and also exacerbate their health, especially for those with underlying health problems. Some lived experiences of residents of informal settlements and their daily struggles for food, shelter, access to health services, and their cumulative effect on their lives are documented, reflecting on the SDG targets and indicators. Finally, the authors position the discussion at the national level by looking at the challenges facing the state in meeting these basic needs in the light of dwindling resources. Recommendations are made towards ending the extreme poverty experienced in informal settlements.

Chapter 3 by Maureen Maisiri, Moleboheng B Pherane, Palesa B Ayantokun, Leemisa S Matooane, Mischka Dunn and Abraham Matamanda builds on Chap. 2 which addresses poverty issues related to urban health. Specifically, this chapter examines how the second SDG on ending hunger relates to urban health. In their examination, the authors consider the situation of food access and supply among poor residents in the Dinaweng informal settlement. The prevalence of hunger is shown in the poorer neighbourhoods which has direct impacts on attaining SDG 2 because the strides towards attaining this goal are derailed and the well-being of the residents is compromised as they skip meals, eat paltry portions, lack an adequate and balanced diet; hence their bodies are not well nourished. Poor neighbourhoods are characterized as food deserts as the residents must resort to unhealthy foods. Some rely on informal food vendors who products may not always be good for their well-being. Hence, concerted efforts need to be made to alleviate hunger as this directly impacts the health of the communities.

Chapter 4 is by Abongile Mgwele, Johannes Bhanye and Kgosi Mocwagae. This chapter touches on a critical issue of water and sanitation and the link to urban health. Adopting a feminist perspective, the authors assert the argument raised by Tulchinsky (2018) that waterborne disease is given less recognition as water and sewerage treatment standards are assumed to be part of modern living. It is found that, while substantial progress has been made in increasing access to clean drinking water and sanitation, billions of people—mostly in developing countries—still lack these basic services. The COVID-19 pandemic has demonstrated the critical importance of sanitation, hygiene and adequate access to clean water for preventing and containing diseases. Women and girls—especially those living in urban poverty—are disproportionately affected by access to water and sanitation. This chapter explores the gender disparities and implications of inadequate water and sanitation in the Mangaung Metropolitan Municipality. Local level coping strategies to water and sanitation challenges among women in selected communities in Mangaung are explored. Data for this study is based on a survey.

Siphokazi Rammile, Johannes Bhanye, Timothy Lehobo and Mareli Hugo extend the discussion presented in Chap. 5. This chapter focuses on SDG 9 and specifically interrogates how innovations and infrastructure correlate with urban health in poor neighbourhoods of Bloemfontein. The authors explore the progress in achieving SDG 9, building resilient infrastructure, and promoting and fostering innovation concerning urban health. The chapter demonstrates that building resilient infrastructure and fostering innovation for urban health in the African city's marginalized communities is urgently required to support progress in the achievement of the SDGs and other global and regional public health targets, including, the new Urban Agenda, and the African Union's Agenda 2063.

In Chap. 6, Lucia Leboto-Khetsi, Jennilee Kohima, Richman T Gambe, Siphokazi Rammile and Thulisile N Mphambukeli deliberate on SDG eleven and how it impacts on urban health. Using a case study of a hostel in Bloemfontein, the authors bring out the intricacies in how the poor reside in squalid conditions akin to "*the city of the dreadful night*" depicted by Peter Hal who recounted the appalling conditions residents faced in their apartments during the Industrial Revolution. The authors argue that the state of the hostel illustrates the failures and negligence of the housing needs of the poor, which eventually places them at a disadvantage as their residence exposes them to multiple health risks. These health risks are contrary to the aspirations of SDG 11, envisaging safe, resilient, sustainable, and inclusive settlements for all by 2030. The authors conclude that are these old hostels present an opportunity to be remodelled into a safe liveable neighbourhood through collaborative efforts by the state, community, private sector, academia, and other relevant stakeholders.

The impacts of climate change are keenly felt in cities. As many proponents have alluded, the effects are most severe for the urban poor, due to their greater vulnerability. SDG 13 represents a commitment to undertake climate action to reduce the effects of climate change. Hence, in Chap. 7, Halleluah Chirisa, Adriaan van der Walt, and Abraham Matamanda situate their contribution to the discourse on climate change and urban health nexus. The chapter aims to understand the link between climate action SDG with social determinants of health, to review and analyze the strides and milestones in implementing the SDG 13 through a policy review in South Africa and the MMM context. The chapter further analyses and assesses local climate actions by MMM communities in dealing with climate change and its impact on health.

In Chap. 8, Simbarashe Jombo argues that healthy cities extend beyond the provision of health infrastructure and primary health care. Instead there is an intricate relationship of a healthy and liveable city and the presence of a healthy natural environment. This chapter contributes to this existing knowledge by examining how the aspirations to protect the environment are espoused in SDG 15 and are articulated in the urban health nexus. It emphasizes that protecting the natural environment contributes to public health among urbanites. The authors employ the socio-ecological systems theory and are premised on the fact that reducing environmental degradation plays an essential role in developing strategies and policymaking towards a sustainable urban environment.

Abraham Matamanda, Fortune Mangara, Tafadzwa Maramura and Charles Mukumba explore how SDG 16 relating to institutions, peace, justice and inclusivity contributes to urban health. In Chap. 9, the authors employ socio-spatial justice and post-colonial theories in understanding how institutions can enable or disable peace and justice, which ultimately affects urban health as some individuals are marginalized and eventually left behind. An inclusive, democratic and peacebuilding approach are emphasized in this chapter, which is suggested as a practical way to foster peaceful and inclusive societies for sustainable development.

In Chap. 10, Kgosi Mocwagae, Innocent Chirisa and Tebello Putsoane explore the lived experiences of energy consumers in Bloemfontein reflecting on the health nexus of the energy sources used by the poor. The chapter uses the lens of Sustainable Development Goals 7 and 12 to explore the options of energy consumers in low-income settlements in Bloemfontein. Many livelihoods strategies in informal settlements revolve around the commoditization of electricity and related products as people seek to charge their cell phones, among other gadgets. The successful attainment of Sustainable Development Goals is challenged mainly by the marginalisation and powerlessness of informal settlement dwellers.

In Chap. 11, Verna Nel and Abraham Matamanda conclude the book by synthesizing the issues raised in the respective chapters.

References

Africa CDC (2024) Africa CDC support on cholera outbreak in Zambia. Available from: https://africacdc.org/news-item/africa-cdc-support-on-cholera-outbreak-in-zambia/#:~:text=About%20the%20Outbreak%20in%20Zambia&text=As%20of%2019%20January%202024,while%20249%20were%20community%20deaths (Accessed 13 Feb 2024)

Ahsan MM (2020) Strategic decisions on urban built environment to pandemics in Turkey: lessons from COVID-19. J Urban Manag 9(3):281–285

Auditor General of South Africa (2018) Mangaung Metropolitan Municipality consolidated annual financial statements for the year ended 30 June 2017. Auditor General of South Africa, Pretoria

Barton H (2009) Land use planning and health and well-being. Land Use Policy 26:115–123

Battersby J, McLachlan M (2013) Urban food insecurity: a neglected public health challenge. S Afr Med J 103(10):716–717

Bingham P, Verlander NQ, Cheal MJ (2004) John Snow, William Farr and the 1849 outbreak of cholera that affected London: a reworking of the data highlights the importance of the water supply. Public Health 118(6):387–394

Brousse O, Georganos S, Demuzere M, Vanhuysse S, Wouters H, Wolff E, Linard C, Nicole PM, Dujardin S (2019) Using local climate zones in sub-Saharan Africa to tackle urban health issues. Urban Clim 1(27):227–242

Campbell K, Cowan R (2016) The radical incrementalist: how to build urban society in 12 lessons. Massive Small

Caplan JM, Kennedy LW, Neudecker CH (2020) Cholera deaths in Soho, London, 1854: risk terrain modeling for epidemiological investigations. PLoS One 15(3):e0230725

Chigudu S (2020) The political life of an epidemic: cholera, crisis and citizenship in Zimbabwe. Cambridge University Press, Cambridge

Corburn J (2009) Toward the healthy city. MIT Press, Cambridge

Corburn J (2013) Healthy city planning. Routledge, London

Dixon J, Omwega AM, Friel S, Burns C, Donati K, Carlisle R (2007) The health equity dimensions of urban food systems. J Urban Health 84(1):118–129

Ford N (2023) The SDGs in Africa—challenges to be overcome. African Business, 20 September. https://african.business/2023/09/resources/the-sdgs-in-africa-challenges-to-be-overcome#:~:text=SDG%201%20calls%20for%20ending,in%20extreme%20poverty%20in%202022 (Accessed 28 Feb 2024)

Gwenzi W, Sanganyado E (2019) Recurrent cholera outbreaks in sub-Saharan Africa: moving beyond epidemiology to understand the environmental reservoirs and drivers. Challenges 10(1):1

Haase D, Güneralp B, Dahiya B, Bai X, Elmqvist T (2018) Global urbanization. In: Elmqvist T, Bai X, Frantzeskaki N, Griffith C, Maddox D, McPhearson T, Parnell S, Romero-Lankao P, Simon D, Watkins M (eds) Urban planet: knowledge towards sustainable cities. Cambridge University Press, Cambridge

Hancock T, Duhl L (1988) WHO healthy cities project: promoting health in the urban context. In: WHO healthy cities papers No. 1. World Health Organization, Geneva

Harris B, Goudge J, Ataguba JE, McIntyre D, Nxumalo N, Jikwana S, Chersich M (2011) Inequities in access to health care in South Africa. J Public Health Policy 32:S102–S123

Jackson LE (2003) The relationship of urban design to human health and condition. Landsc Urban Plan 64(4):191–200

Jones N, Gwata M, Akoon I (2022) Beating the heat in South African cities: lessons from a citizen science assessment. World Bank Blogs, 17 November. Available at https://blogs.worldbank.org/sustainablecities/beating-heat-south-african-cities-lessons-citizen-science-assessment (Accessed 16 Feb 2024)

Kimani-Murage EW, Schofield L, Wekesah F, Mohamed S, Mberu B, Ettarh R, Egondi T, Kyobutungi C, Ezeh A (2014) Vulnerability to food insecurity in urban slums: experiences from Nairobi, Kenya. J Urban Health 91(6):1098–1113

Kjellstrom T (2016) Impact of climate conditions on occupational health and related economic losses: a new feature of global and urban health in the context of climate change. Asia Pac J Public Health 28(2_suppl):28S–37S

Kroll C, Warchold A, Pradhan P (2019) Sustainable development goals (SDGs): are we successful in turning trade-offs into synergies? Palgrave Commun 5(1):1–11

Kuddus MA, Tynan E, McBryde E (2020) Urbanization: a problem for the rich and the poor? Public Health Rev 41(1):1–4

Labuschagne I (2022) South Africa's hunger problem is turning into a major health crisis. The Conversation, 1 June. Available at: https://theconversation.com/south-africas-hunger-problem-is-turning-into-a-major-health-crisis-183736 (Accessed 16 Feb 2024)

Leal Filho W, Wolf F, Salvia AL, Beynaghi A, Shulla K, Kovaleva M, Vasconcelos CRP (2020) Heading towards an unsustainable world: some of the implications of not achieving the SDGs. Discov Sustain 1(2). https://doi.org/10.1007/s43621-020-00002-x

Lenka M, Marais L, Rani K, Tsotetsi LD (2021) Mangaung State of the Cities report—South African Cities Network. Johannesburg

Mabaso ML, Vounatsou P, Midzi S, Da Silva J, Smith T (2006) Spatio-temporal analysis of the role of climate in inter-annual variation of malaria incidence in Zimbabwe. Int J Health Geogr 5:1–9

Manning G (2021) Hunger in South Africa and solutions. The Borgen Project Blog. https://borgenproject.org/hunger-in-south-africa/ (Accessed 16 Feb 2024)

Marais L, Twala C (2021) Bloemfontein: the rise and fall of South Africa's judicial capital. Afr Geogr Rev 40(1):49–62

Matamanda AR, Nel V (2021) (Un)healthy cities: reflections on urban public health in poor neighbourhoods. In: Matamanda AR, Nel V, Chirisa I (eds) Urban geography in postcolonial Zimbabwe: paradigms and perspectives for sustainable urban planning and governance. Springer Nature, Cham, pp 93–118

Matamanda AR, Dunn M, Nel V (2022a) Broken bridges over troubled waters: COVID-19 and the urban poor residing in Dinaweng informal settlement, Bloemfontein, South Africa. S Afr Geogr J 104(3):309–327

Matamanda AR, Nel V, Dunn M, Mgwele A, Rammile S, Leboto-Khetsi L, Kohima JM, Ngo PB (2022b) Sustainable development goals and urban health challenges in informal settlements of Mangaung Municipality, South Africa. In: Filho WL, Azul AM, Doni F, Salvia AL (eds) Handbook of Sustainability Science in the Future: policies, Technologies and Education by 2050. Springer Nature, Cham. https://doi.org/10.1007/978-3-030-68074-9

Matamanda AR, Mphambukeli TN, Chirisa I (2024) Exploring water-gender-health nexus in human settlements in Hopley, Harare. Cities & Health 8(1):44–58

Mathaai W (2009) The challenge for Africa. Pantheon Press

Mengel MA, Delrieu I, Heyerdahl L, Gessner BD (2014) Cholera oubreaks in Africa. In: Nair G, Takeda Y (eds) Cholera outbreaks. Current Topics in Microbiology and Immunology, vol 379. Springer, Berlin, pp 117–144

Mentz L (2024) Heat and more heat for the next few days in Bloemfontein. News24, 12 February (Accessed 02 Mar 2024)

Mumford L (1961) The city in history: its origins, its transformations, and its prospects. Harcourt, Brace, New York

Patz JA, Grabow ML, Limaye VS (2014) When it rains, it pours: future climate extremes and health. Ann Glob Health 80(4):332–344

Pinchoff J, Mills CW, Balk D (2020) Urbanization and health: the effects of the built environment on chronic disease risk factors among women in Tanzania. PLoS One 15(11)

Ramin BM, McMichael AJ (2009) Climate change and health in sub-Saharan Africa: a case-based perspective. EcoHealth 6(1):52–57

Republic of South Africa (1996) Constitution of the Republic of South Africa. Republic of South Africa, Pretoria

Semela B (2023) One child dead and several people injured in Bloemfontein storm. GroundUp, 8 December. https://groundup.org.za/article/one-child-died-and-several-people-injured-in-devastating-bloemfontein-storm/ (Accessed 02 Mar 2024)

Stats SA (2022) Provinces at a glance. Stats SA, Pretoria

Subramanyam N, Marais L (2022) Making Mangaung Metro: the politics of metropolitan reform in a South African Secondary city. Urban Stud 59(14):2893–2911

Tacoli C (2020) Food (in) security in rapidly urbanizing, low-income contexts. In: Crush J, Frayne B, Haysom G (eds) Handbook on Urban food security in the global south. Edward Elgar Publishing, pp 23–33

Tulchinsky TH (2018) John Snow, Cholera, the broad street pump; Waterborne diseases then and now. Case Stud Public Health 2018:77–99

Viglione G (2023) In-depth: how climate change affects health in Africa. Carbon Brief

Vlahov D, Freudenberg N, Proietti F, Ompad D, Quinn A, Nandi V, Galea S (2007) Urban as a determinant of health. J Urban Health 84(1):16–26

Wilkinson A (2020) Local response in health emergencies: key considerations for addressing the COVID-19 pandemic in informal urban settlements. Environ Urban 32(2):503–522

2 Ending Extreme Poverty and Enhancing Urban Health

Natal Buthelezi, Lucia Leboto-Khetsi, and Verna Nel

2.1 Introduction

Cities are the cores of economic development, improved infrastructure and services, healthcare, education, social inclusion and global competitiveness (Côté-Roy and Moser 2019; Förster and Siegenthaler 2018). However, alongside the promise of urban prosperity lies the stark reality of deep-rooted challenges, particularly in complex contexts of Sub-Saharan Africa (Finn and Cobbinah 2023; Li et al. 2022). As urbanization surges with the influx of people, creating healthy and inclusive cities becomes imperative to address the pressing issues of extreme poverty (Bodo 2019; Kundu and Pandey 2020; Nzimakwe 2020).

Extreme poverty, as defined by the Sustainable Development Goals (SDGs), denotes the most severe and dire manifestation of impoverishment, wherein individuals and households endure conditions marked by extreme deprivation, insecurity and a lack of access to basic human needs (United Nations 2022). The SDGs recognize that poverty is complex, multi-dimensional and encompasses a range of interconnected dimensions, even though living on less than US$2.15 per day (revised from the US$1.90 poverty threshold in 2015), remains a crucial aspect of measuring extreme poverty (Carr 2021; Fransman and Yu 2019). Therefore, it is essential to acknowledge how a particular setting has a significant impact on how poverty emerges, presenting a complex and ambiguous challenge (Matamanda et al. 2022; Ngepah et al. 2023; Turok and Borel-Saladin 2018). Nevertheless, it is precisely this multi-faceted nature and ambiguity that prompted global leaders to prioritize poverty eradication in the SDGs. The SDG1 underscores the imperative to eradicate poverty in all its forms and dimensions, acknowledging its foundational role in achieving broader sustainable development objectives (Bazilah et al. 2020; World Bank 2020b).

The importance of addressing extreme poverty on a global scale cannot be overstated. Beyond personal struggles, poverty is a systemic problem that perpetuates inequality, societal discontent, impedes health and well-being, and hampers the achievement of broader SDGs (Carr 2021; Francis and Webster 2019; Satterthwaite et al. 2019). Often, poverty ensnares both indi-

N. Buthelezi (✉)
Department of Urban & Regional Planning, University of the Free State, Bloemfontein, South Africa

L. Leboto-Khetsi
Center for Development Support, University of the Free State, Bloemfontein, South Africa

V. Nel
Department of Urban & Regional Planning, University of the Free State, Bloemfontein, South Africa

Center for Development Support, University of the Free State, Bloemfontein, South Africa

A. R. Matamanda, V. Nel (eds.), *Sustainable Development Goals and Urban Health*, Sustainable Development Goals Series, https://doi.org/10.1007/978-3-031-68734-1_2

viduals and communities in a downward spiral, limiting opportunities for prosperity and jeopardizing basic dignity (Chirisa and Matamanda 2016; Schmidt 2010). This chapter delves into the nexus between urban health and poverty, examining how the complexities of urban living can exacerbate inequalities and contribute to the persistence of poverty in South Africa's urban areas, including Bloemfontein.

Bloemfontein in the Free State province of South Africa, epitomizes the complexities of urbanization in a developing country (Marais 2021; Mphambukeli 2019). The city's population experienced a significant increase due to urban development and an influx of migrants in search of improved prospects, resulting in an annual growth rate of 1.76%, increasing from 556,450 individuals in 2019 to 589,383 by 2023 (Manji et al. 2023; Mphambukeli and Nel 2018; World Population Review 2023). Subramanyam and Marais, (2022) illustrated that this influx has strained existing infrastructure, services and resources, exposing the disparities, which persist within its boundaries. The persistence of extreme poverty lies heart of Bloemfontein's challenges (Mphambukeli 2019). Despite notable economic growth averaging 3.32% annually between 2004 to 2014 and 2.01% between 2014 to 2019, the city continues to grapple with high poverty rates (CoGTA 2020). Around 61% of all households earn below R3500 (approximately US$187) per month, with the majority of the population struggling to access basic amenities, education and healthcare (Mangaung Metropolitan Municipality 2021; Verster 2018). This stark juxtaposition of prosperity and deprivation underscores the need for comprehensive and context-sensitive efforts to address the root causes of poverty and inequity.

The study's theoretical framework is based on the determinants of health and poverty which acknowledge the complex interplay of numerous elements, including social, economic, political and environmental factors. By applying this framework, the paper examines how spatial inequalities, economic disparities and environmental conditions in Bloemfontein contribute to urban health disparities and perpetuate the cycle of poverty. This theoretical lens provides a comprehensive understanding of the complex relationship between poverty and urban health, emphasizing the importance of addressing these diverse determinants to achieve sustainable improvements in health and inequality disparities.

Addressing extreme poverty and advancing urban health aligns with the SDGs outlined by the United Nations. These global aspirations find resonance within the context of South Africa's National Development Plan, which envisions a future characterized by equality, inclusivity and improved quality of life for its citizens (Cumming et al. 2017; Zarenda 2013). However, this chapter argues that the task requires a comprehensive strategy, which intertwines spatial planning with principles of health equity and inclusivity. This involves collaborative efforts between sectors, community engagements in the planning process and the implementation of policies prioritizing equitable distribution and opportunities for all residents. The case study of Phomolong serves as a microcosm of these efforts, offering a lens to scrutinize their effectiveness. By examining the success and gaps in interventions across various sectors, we gain valuable insights into the real-world implications of this integrated approach, ultimately contributing to the shared pursuit of SDG1, SDG3 and South Africa's development aspirations.

This chapter begins with a literature review that defines the concept of poverty and discusses the relevance of SDGs, notably SDG1, in addressing extreme poverty and its impact on urban health. It then transitions to examine specific poverty challenges in post-apartheid South Africa. The research methodology is outlined, followed by the presentation of research findings. A discussion section synthesizes empirical data with theoretical insights and the chapter concludes by summarizing the key elements of the research.

2.2 Literature Review

2.2.1 Concept of Poverty

Poverty is a complex and multifaceted issue that encompasses both measurable, quantifiable indicators (objective) and individual perceptions or experiences (Azevedo 2020a, b; Fransman and Yu 2019; Qizilbash 1998). In monetary terms, poverty is associated with quantifiable factors about income and affordability of essential items for survival. They include poverty lines, income distribution analysis, and other economic indicators (World Bank 2020a). For instance, there are considerable regional differences in poverty lines, which serve as a benchmark for determining the minimum income required to meet basic needs. Between 2018 and 2019, 41% of the population in Sub-Saharan Africa lived below the US$2.15 per day international poverty threshold. The rate of extreme poverty is 54.8% in central Africa and 45.1% in southern Africa. In western and eastern Africa, poverty rates are 36.8% and 33.8% respectively (Aikins and Mclachlan 2022; World Bank 2020a).

Non-monetary poverty includes inadequate access to public services, limited asset ownership, social isolation, low educational attainment, poor health and vulnerability to crime (World Bank 2019). These measures recognize that poverty is not solely about income but also about lacking essential resources and opportunities. The subjective aspects of poverty recognize the importance of understanding people's experiences, aspirations and psychological well-being (Azevedo 2020a, 2020b; Castaneda et al. 2016). Although no one definition of poverty is universally accepted, it comprises limited essential opportunities and choices, which are necessary for human development and a decent standard of living (Jansen et al. 2015). Therefore, a clear and precise definition of poverty is essential to understand its causes, gauge its severity and evaluate progress made towards its eradication.

In South Africa, households' poverty status fluctuates over time, making it a dynamic phenomenon rather than a static state (World Bank 2020a). This emphasizes the importance of understanding the multiple dimensions of poverty and the different forms it takes in the country's historical context (Graven 2014; World Bank 2020b). In the past, poverty was often shaped not only by income levels but also by systemic denial of access to basic services and opportunities for specific racial groups (Bhorat and Kanbur 2006). This chapter, therefore, argues that measuring poverty in financial terms might only partially capture the complexities thereof in South Africa. Broader definitions that consider aspects of human development and well-being, can provide a more comprehensive understanding of poverty in the country.

2.2.2 Sustainable Development Goals and the Interplay Between Ending Extreme Poverty and Urban Health

The SDGs address complex issues that face humanity, with SDG1 as a cornerstone to eradicate extreme poverty in all its forms by the year 2030. Simultaneously, SDG3 emphasizes the promotion of health and well-being, acknowledging the interplay among the SGDs. Several health targets are contained in SDG3, including reducing maternal and infant mortality, preventing communicable diseases, promoting mental health and achieving universal health coverage to ensure the well-being and health of all people, regardless of age and gender (United Nations 2022). However, the achievement of SDG1and SDG3 are connected to the successes of other related goals (Fonseca et al. 2020; Nhamo et al. 2020; Pradhan et al. 2017). For instance, universal access to safe and affordable housing, basic services, upgrading of slums and affordable transport systems (SDG11), provision of clean water and sanitation (SDG6), resilient infrastructure development (SDG9), climate action (SDG13), ending hunger (SDG2) and decent work opportunities irrespective of age, education and gender

(SDG8), all play essential roles in ensuring the realisation of SDG1 (Fonseca et al. 2020). The relationship between SDG1 (eradicating poverty) and SDG3 is particularly crucial, because both share the goal of improving human well-being and promoting sustainable development (Lavell et al. 2023).

However, while these targets are commendable, achieving poverty reduction in Sub-Saharan Africa presents unique challenges (Ayoo 2022; Mubecua and David 2019). The region faces a combination of prevalent poverty, high inequality and slow income growth, making the task more daunting (Chirisa and Matamanda 2016; Kwasi Fosu 2018). Beegle and Christiaensen (2019) and Fofana et al. (2023) state that extreme poverty will become increasingly concentrated in Africa compared to other regions of the world. The complexity of this phenomenon is evident in the Sub-Saharan urban landscape with the emergence of densely populated slums and informal settlements, which concentrates the impoverished population and highlights stark health disparities (Huchzermeyer 2019; Oksiutycz and Azionya 2022; Richmond et al. 2018). Residents in places like Kibera in Nairobi (Kenya), Hopley farm in Harare (Zimbabwe) and Duncan Village in East London (South Africa) live under harsh circumstances with a lack of basic amenities (water, healthcare, job opportunities, security and proper physical infrastructure), subsequently exacerbating health disparities and limiting access to essential services (Kim et al. 2022; Matamanda 2024; Ndasana et al. 2022; Turok and Borel-Saladin 2018).

Therefore, it is crucial to address the connected problems of poverty and health to promote sustainable growth within the dynamic environment of urbanization. This section explores the critical intersections between the SDGs. Ending and reducing extreme poverty and enhancing urban health is analysed, which sets a standard for worldwide efforts and highlights the urgency and significance of addressing this complex issue.

2.2.3 The Poverty Conundrum in Post-apartheid South Africa

Poverty in post-apartheid South Africa is a deeply entrenched issue with clear racial, gender and spatial dimensions, stemming from the historical policies of colonial, segregationist and apartheid regimes (Triegaardt 2006). Despite the country's middle–income status, extreme levels of inequality persist, with a vast number of people living in abject poverty while a smaller elite enjoys immense wealth (World Bank 2020a). This inequality is apparent in the stark contrast in urban spatial development, characterized by informal settlements, slum conditions, and inadequate basic services of the many versus the opulence of a privileged few (Triegaardt 2006; Turok and Borel-Saladin 2018; Verster 2018).

According to Stats SA (2021:4), South Africa had a substantial upward shift from 2006 to 2021 in the inflation-adjusted poverty line, increasing from R219 (approximately US$31) to R624 (approximately US$39) per person per month. Similarly, the lower bound poverty line escalated from R370 (US$53) to R890 (US$56), while the upper bound poverty line experienced a significant growth from R575 (US$82) to R1335 (US$84). This suggests that the cost of meeting basic needs has increased considerably. Increases in food poverty lines highlight that affordable food has become more challenging for a significant portion of the population. The escalation in the lower and upper-bound poverty lines further underscores the overall economic strain experienced by households, with more people potentially struggling to afford not only food but other essentials (Jansen et al. 2015; World Bank 2020a). The widening gap between the poverty lines further indicates a growing disparity in income distribution (Burns et al. 2017; Francis et al. 2020). The fact that the upper-bound poverty line increased more than the lower-bound poverty line implied that economic challenges have disproportionately affected those with lower incomes (Francis and Webster 2019; Hakizimana and Geyer 2014).

The eradication of poverty in South Africa is a crucial social, political and economic imperative as it adversely affects human well-being and sustainable livelihoods (Fransman and Yu 2019; Triegaardt 2006). Interventions and pro-poor policies aimed at poverty alleviation, including social grants, housing initiatives and healthcare reforms have provided some relief for millions of South Africans living in poverty. However, economic growth has stagnated, inflation remains high, and the unemployment rate has risen, leading to a reversal of previous declines in poverty rates (Stats SA 2021).

Gbadegesin et al. (2020), Mphambukeli (2019) and Marutlulle (2021) argue that addressing poverty in South Africa necessitates accurate identification of the most deprived areas and targeted implementation of appropriate poverty-reduction strategies. The country's progress in combating poverty and inequality is tied to its constitution, which emphasizes the advancement of socio-economic rights (Langford et al. 2013; McLean 2009; Pieterse 2004). Francis (2006) and Ramphoma (2014) argue that to tackle the poverty conundrum, South Africa must adopt a comprehensive approach, that addresses the root causes of poverty including systemic discrimination and unequal distribution and resources. Economic programmes such as the National Development Plan, Broad-Based Black Economic Empowerment (B-BBEE) and Spatial Planning and Land Use Management Act (SPLUMA) have been implemented to redress historical imbalances, but more needs to be done to create inclusive growth and opportunity (Fourie 2018; Juggernath et al. 2011; Nel 2016).

2.3 Research Methodology

This qualitative case study centred on the Phomolong informal settlement in Bloemfontein, South Africa, is a focused-on poverty and health dynamics. In-depth interviews were the primary data collection methods, offering insights into the multifaceted aspects of poverty in Phomolong, particularly concerning housing conditions, local economics, service delivery and health. After ethical clearance was obtained, interviews were conducted with 26 randomly sampled respondents, which included representatives from across the settlement. The study explored respondents' perspectives and experiences, facilitating a comprehensive understanding of poverty-related health indicators within the settlement. Additionally, secondary data sources were incorporated into the research. The secondary data sources from reputable institutions such as Stats SA, Mangaung Metropolitan Municipality, and the Department of Cooperate Governance and Traditional Affairs (CoGTA) provided a comprehensive context and statistical foundation to assess poverty and well-being trends in the area. Thematic analysis was used to identify patterns and narratives in the qualitative data. Socioeconomic vulnerability and housing conditions, health disparities, community resilience and coping mechanisms, barriers in accessing health services and structural inequalities with policy implications were among the subjects covered.

2.4 Findings

2.4.1 National, Provincial and Local Government Response Strategies to SDG1

Since the Reconstruction and Development Programme[1] (RDP) in 1994, which laid the foundation for the post-apartheid government to reduce poverty in all its manifestations, South Africa has implemented a variety of policy formulation strategies to eradicate poverty and inequality. The RDP has since become the

[1] RDP is South Africa's socioeconomic policy framework established during Nelson Mandela's presidency in 1994. Its primary objective was to eradicate inequality by means of reconstruction, development and the creation of an inclusive democratic society free from racial and gender discrimination.

National Development Plan (NDP), which outlines a long-term vision for combating poverty, inequality and unemployment by the year 2030 (Masilela and Mthiyane 2014). There is a strong focus on enhancing economic growth by improving infrastructure development, public service, education and skills, implementing spatial development measures and providing assistance to small businesses (National Planning Commission 2012). However, it appears that the idealized South Africa envisioned by 2030 may not materialize as initially hoped given the missed milestones listed in the policy (Fourie 2021). South Africa has struggled with growing unemployment, pervasive corruption, overburdened education and healthcare systems – the after-effects of apartheid and challenging external environments – which contributed to weak economic growth (International Monetary Fund 2023; Mackett 2020).

The Social Assistance Act 16 of 2020 establishes the foundation for social grants and assistance programmes, offering support to vulnerable individuals and families (Republic of South Africa 2020). Grants have played an important role in the fight against poverty in South Africa by serving as a safety net for individuals who cannot work for pay, children, pensioners and persons with disabilities, all of whom need better living conditions (Mackett 2020). Social grants have been credited for their effectiveness in addressing the needs of vulnerable groups, alleviating immediate problems and creating a foundation for long-term development (Gutura 2014; Schiel et al. 2016). According to Stats SA (2021), grants were the primary source of income for nearly one-fifth (24%) of households nationwide, coming in second to salaries (59.4%) in importance. More households received grants compared to salaries as a source of income in Mpumalanga (66.2% versus 50.9%), Limpopo (65.7% versus 49.7%), Eastern Cape (63.7% versus 46.2%), and Free State (60.0% versus 53.2%). However, despite the positive impacts of social grants on poverty eradication, challenges persist. The sustainability of grants programmes, administrative efficiency and preventing misuse or fraud remain ongoing concerns (Mashaba and Saurombe 2023).

Nationally, progress towards achieving SDG1 has been assessed by specific targets and commitments, illustrated in Table 2.1. The implementation of the minimum wage act in response to eradicating extreme poverty for all people represents a significant step in this direction. However, it is noteworthy that the percentage of people below the poverty line has seen a slight increase, rising from 25.4% in 2006 to 26.7% in 2022 (Cowling 2023). This indicates that a major challenge persists in the endeavour to reduce poverty.

Another key target of SDG1 is to ensure equal rights to economic resources, access to basic services and tenure security by 2030. This target is being pursued nationally through initiatives such as building human settlements in areas that offer economic opportunities, and providing essential services and amenities (National Planning Commission 2012). The National Housing Policy offers guidelines for housing delivery underpinned by principles of sustainability, viability, integration, equity, reconstruction, holistic development and good governance (Department of Human Settlements 2015). As part of the state's efforts to improve housing and living conditions in informal settlements, the strategy of *in situ* upgrading was also established (Huchzermeyer 2009). Additionally, legal frameworks like the Local Government: Municipal Systems Act 32 of 2000 and SPLUMA guide governance, service delivery and land use management, locally and nationally. The frameworks have been formulated to meet the set targets for the NDP and SDGs. Notable progress has been made, as demonstrated by a 2% improvement in sanitation from 2015 to 2017, as well as a 0.9% increase in access to electricity during the same period. However, there was a decrease in the provision of water supply by 1.1% from 2015 to 2017 (Stats SA 2021).

Efforts are underway to build resilience for the poor and reduce vulnerability to climate change.

Table 2.1 Evaluating Sustainable Development Goal (SDG) targets and Republic of South Africa's priorities and commitments

SDG	Target	National commitment	Remarks
Goal 1: No Poverty	By 2030 eradicate extreme poverty for all people everywhere.	Implementation of the Minimum Wage Act.	The number of people below the poverty line increased from 25.4% (2006) to 26.73% (2022). Major challenges remain. (Stats SA 2021)
	By 2030 have equal rights to economic sources, access to basic services, and tenure security.	Building human settlements in areas that bring economic opportunities and all the services and amenities people need. Allocate serviced stands.	Improved sanitation by 2% from 2015–2017. Improved electricity by 0.9% from 2015–2017. Reduced provision of water supply by 1.1% from 2015–17. (Stats SA 2021)
	By 2030 build resilience for the poor and reduce vulnerability to climate change.	Implementing a low-carbon transition. Better land use planning and compact city development. Periodic risk assessments of human settlements.	Increased number of deaths caused by local disasters. 13 disaster risk reduction strategies adopted.
	Create pro-poor and gender-sensitive development strategies.	Protect and enhance the environmental assets and natural resources.	Poor execution of strategies, little coordination across initiatives, no prioritisation and inadequate attention to resourcing.
	Ensure equal opportunity and reduce inequalities of outcome.	Promote the Bill of Rights.	Improvements have been recorded.

This involves implementing a low-carbon transition, adopting better land use planning and promoting compact city development while conducting periodic risk assessments in human settlements (Huxham et al. 2019; Nel 2011). It is important to note, however, that there was an increase in deaths caused by local disasters, underscoring the need for effective disaster risk reduction strategies, with 13 such strategies having been adopted (Bond and Galvin 2022; Mazeka et al. 2019; Xulu et al. 2020).

There is a focus on creating development strategies that are pro-poor and gender-sensitive, as well as protecting and enhancing environmental assets and natural resources (Gunter 2020; Jili et al. 2019; van der Bank and Karsten 2020). Unfortunately, challenges persist in the execution of these strategies, with limited coordination across initiatives, a lack of prioritization and insufficient attention to resource allocation (Gunter and Massey 2020). Progress has been made in ensuring equal opportunities and reducing inequalities of outcomes, with improvements noted in the promotion of the bill of Rights. Nevertheless, there remains room for further advancements in this regard (Francis and Webster 2019).

2.4.2 Experiences from Phomolong Informal Settlement

This section details the local experiences in an informal settlement. It aims to evaluate how the state's response strategies to SDG1 have reached the most vulnerable individuals within the Bloemfontein urban community.

2.4.2.1 Building Typology and Tenure Security

Ninety-five percent of the respondents in Phomolong informal settlement are recognised as

Fig. 2.1 Building typology in Phomolong informal settlement. (Source: Author 2023)

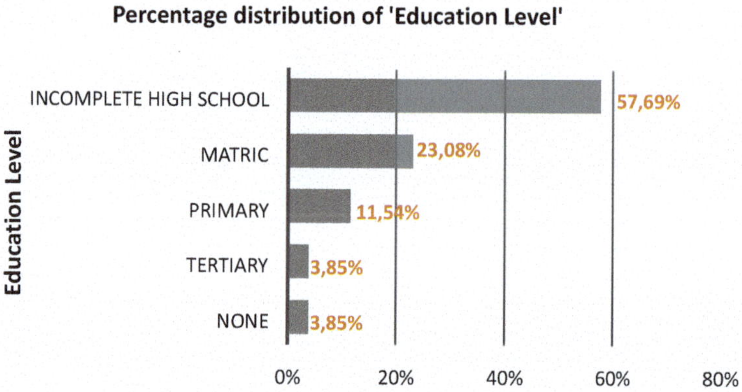

Fig. 2.2 Educational levels of respondents in Phomolong

owners of the shacks, while about 5% are tenants. However, this seemingly positive aspect was overshadowed by the stark reality that homeownership did not guarantee secure tenure in this informal settlement. Residents expressed deep distress and uncertainty regarding their housing situation. The housing landscape in Phomolong is characterised by houses built using corrugated iron sheets, cardboard, planks and other composite waste materials, illustrated by Fig. 2.1. These makeshift dwellings were not well suited to withstand the extreme temperatures experienced during the harsh summer and winter seasons.

As a result, residents faced serious health issues related to poor living conditions. Extreme weather exposes people to increased health risks, such as cold and heat related illnesses in summer and winter, which exacerbates health disparities in urban areas. The respondents also expressed their concerns on the vulnerability of their homes during adverse weather conditions. Natural elements not only posed immediate risks to the residents' safety and well-being but also induced stress and anxiety as they dealt with the danger of property damage and ensuing health effects.

2.4.2.2 Education Levels

Figure 2.2 reveals a distinct trend in the respondents' educational backgrounds, which links to the comprehension of socioeconomic health disparities in this region. More than half of the respondents (57.7%) did not complete high school, followed by 23% successfully finishing their senior secondary education (matric), and 11.5% completed their primary education. However, the proportion of those who completed tertiary education and those who had never received any formal education was low, with both at 3.9%. The distribution of educational levels in Phomolong highlights the prevalence of limited formal education, which can be a barrier to socioeconomic progress and improved health outcomes. Higher education is often associated with better employment opportunities, increased income and a deeper awareness of health-related concerns. Conversely, lower levels of education

Fig. 2.3 Income generating activities in Phomolong. (In South African Townships, Spazas are small convenience shops that offer everyday household products. They are usually run from homes and serve as a source of income for families. These shops are often owned by foreign nationals and are an informal but popular shopping option within the community)

observed in the community can contribute to poverty and limited access to healthcare resources. Thus, people are more susceptible to health hazards and less equipped to make informed health related decisions.

2.4.2.3 Employment

The findings reveal a multi-faceted landscape of employment and income-generating activities within the community, which has a significant impact on the health and well-being of its residents. The majority (46.2%) of the respondents were unemployed. The community's high unemployment rate is noticeable and has profound implications for the residents' socioeconomic and health conditions. Conversely, a notable portion of respondents (26.9%) were engaged in part-time income-generating activities such as recycling, yard cleaning and laundry services. Some residents were involved in informal business activities including selling fruits, vegetables and surplus produce from gardening as illustrated in Fig. 2.3.

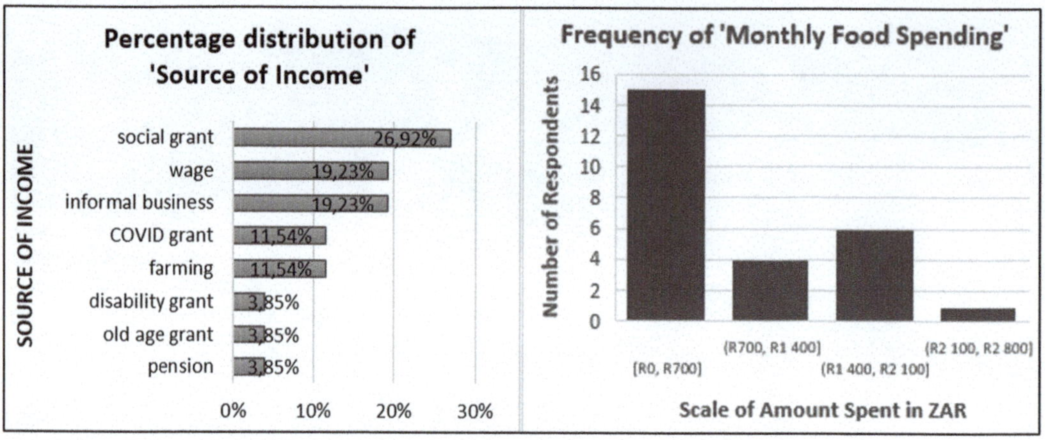

Fig. 2.4 Sources of income and monthly spending on food by Phomolong respondents

Fig. 2.5 Nutritional Garden as a means of food security

These initiatives highlight the community's wide range of economic endeavours as well as the residents' resourcefulness and entrepreneurship skills. Some respondents were formally employed but made up a small portion of the community. The high levels of unemployment in Phomolong can be linked to the limited education attainment, particularly matric and tertiary education. The absence of formal employment opportunities and the prevalence of part-time jobs and other sources of income both contribute to the cycle of poverty and consequently, impact the overall health of the community.

2.4.2.4 Monthly Household Income and Dependence on Grants

Given the high levels of unemployment in Phomolong, it comes as no surprise that most of the respondents relied on social grants as their primary source of household income. Within this context, several types of grants were identified, encompassing disability, old age, COVID-19 and general social grants.

For those who were employed, wages constituted the main source of income, while others generated income through informal business or agricultural activities as illustrated in Fig. 2.4. It is worth noting that some respondents with informal incomes also received periodic social grants as an additional income source. The monthly household income among the respondents exhibited a broad spectrum, ranging from R600 (US$35) to R8000 (US$471) per month. The most common income bracket among respondents fell within the R1000 (US$59) to R1500 (US$88) per month range.

2.4.2.5 Food Security

Each month, respondents spent anywhere from R300 (US$35) to R2800 (US$165) on food. Most respondents spent R0 to R700 (US$41), followed by R140 (US$8) to R2100 (US$124) per month, as illustrated in Fig. 2.4. The least respondents spent between R2100 (US$124) and R2800 (165) per month. Most reported that they frequently ran out of money to buy food to last them for the entire month. The Bloemfontein central business district, community spaza stores, gardening (Fig. 2.5), neighbouring supermarkets and sharing between families provided most of the food.

Respondents relied on various means to secure food, including borrowing from neighbours, obtaining handovers from family members and working in exchange for food. The daily dietary patterns primarily centred on cereal, milk, sugar, vegetables, eggs and fats, with a hierarchical order of consumption. Consumption of roots and tubers, meat and poultry were relatively low, while fruits, eggs and pulses were the least consumed food items. Furthermore, many respondents expressed a desire for changes in their diet to make it healthier and more satisfying. Some mentioned intentions like "adding more nutrients to their household meals," incorporating fruits, increasing meat consumption and reducing the intake of *fast foods* for the sake of their health. Overall, respondents expressed a desire to introduce more variety into their diets.

Remarkably, only five out of the 26 respondents reported that their daily meals provided adequate sustenance throughout the day. When considering annual sustenance, many respondents faced food shortages in January and February due to increased school expenses and during the winter months of June and July. While a few respondents experienced year-round or six-month long food shortages, most encountered food shortages for at least three months within a year. These shortages were attributed to unemployment, financial constraints, poor business performance, insufficient agricultural produce and delays in the disbursements of social grants.

The respondents in Phomolong employed various tactics to address food insecurity on a weekly basis as indicated in Fig. 2.6. Strategies encom-

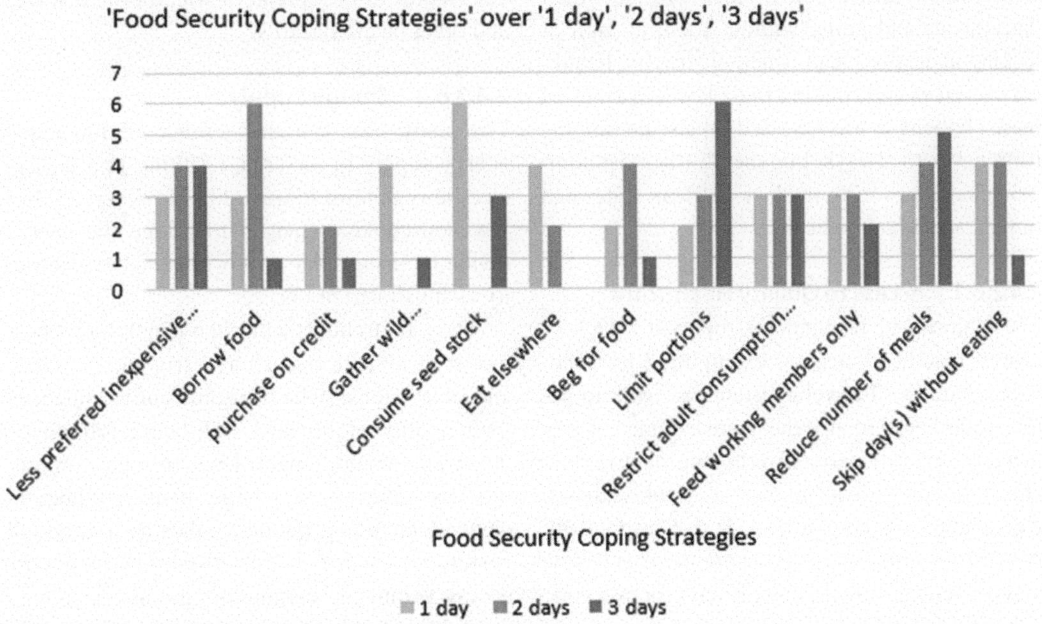

Fig. 2.6 Respondents' food coping strategies in Phomolong

passed actions such as purchasing less preferred cost-effective food items, seeking food through borrowing or begging, reducing meal sizes and prioritizing children's meals over those of adults. Additionally, they engaged in activities like gathering, harvesting or hunting for food, consuming seed reserves, curtailing meal frequency or obtaining food on credit from local stores particularly Spazas. The data revealed that most respondents utilized at least one food security coping strategy on a weekly basis, with the most common being buying less preferred inexpensive food, reducing the number of meals, skipping meals on certain days and consuming seed reserves. Borrowing food occurred at least twice a week, and limiting food portions was practiced for three days a week by most respondents. Generally, some strategy was used in all days of the week, though there was no case of a single respondent having to use a strategy in all days of the week.

2.4.2.6 Poverty and Urban Health Related Challenges in Phomolong

The respondents encountered a range of challenges associated with poverty, which influenced their well-being. Challenges encompassed experiences such as insufficient income to meet their daily needs and going without food, as well as lacking access to clean water, electricity, healthcare services and cooking fuel. The frequency of these challenges was assessed on a scale ranging from never to always. In general, the most common challenges were the lack of clean water and limited access to medical services.

2.4.2.6.1 Access to Quality Healthcare

The majority of respondents reported a lack of nearby healthcare access, exemplified by statements such as, "I travel a distance of ten kilometres when I need to visit the nearest clinic". A few mentioned the occasional presence of a mobile clinic in the township on a weekly basis. Nevertheless, as emphasized by one respondent, "the mobile clinics that operate in the area follow specific schedule, making it difficult for everyone to benefit from them."

2.4.2.6.2 Access to Water

Several respondents indicated that the area had limited access to water supply, with only two having access to private taps equipped with water storage tanks. The rest of the respondents relied on a communal tap as illustrated in Fig. 2.7. They noted that on occasion, the water quality was subpar and there were instances when the water supply was inadequate. One respondent highlighted, "At times, there is a complete lack of water, forcing us to fetch it from the nearest community."

2.4.2.6.3 Sanitation

Respondents of Phomolong predominantly relied on ordinary and ventilated pit latrines for their sanitation needs and male urinals were commonly used as illustrated in Fig. 2.8. One significant issue observed was the disposal of wastewater, which residents typically channelled into holes dug in their yards. However, the presence of open pits posed notable dangers, particularly to children and animals. The precarious nature of these sanitation facilities not only compromised hygiene and safety but also underscored the health disparities faced by the community. The reliance on pit latrines and open wastewater disposal highlights the inadequate sanitation infrastructure in Phomolong, which contributes to the spread of waterborne diseases and other health hazards.

2.4.2.6.4 Energy Supply

The study observed the absence of municipal energy supply in the area, resulting in a lack of electricity in most households. However, a select few households managed to bridge the energy divide by using solar panels to generate electricity as illustrated in Fig. 2.9.

These households with solar panels had access to electrical devices such as refrigerators, radios and televisions, which significantly enhanced their quality of life and well-being. Moreover, they took on additional roles within the community by offering to charge their neighbours' devices, including phones, either as a goodwill gesture or for a fee. The disparities in energy access within the community are striking, with some households enjoying the benefits of elec-

Fig. 2.7 Respondents' water sources in Phomolong

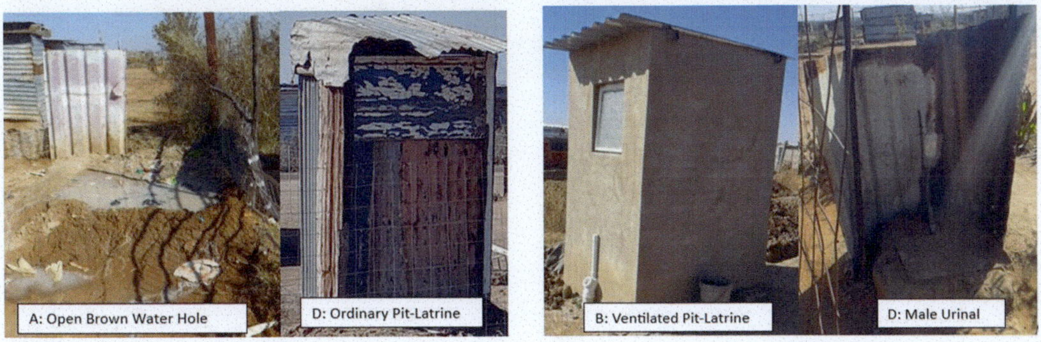

Fig. 2.8 Sanitation provision in Phomolong

Fig. 2.9 Installation of solar panels in Phomolong

Fig. 2.10 Road infrastructure services in Phomolong

tricity while others are left without this essential resource.

2.4.2.6.5 Road Infrastructure

The streets within the locality are composed of gravel and tend to become muddy in rainy seasons, posing challenges for smooth navigation as depicted in Fig. 2.10. Footpaths are used by the majority of the community's inhabitants as means of connectivity. Underdeveloped transportation infrastructure in the area not only isolates residents from job opportunities, education and essential services, but also perpetuates spatial disparities and exacerbates the cycle of poverty.

2.5 Discussion

The results of the study highlight the poverty and urban health conditions of Bloemfontein. Like many other African cities, it is a complex and deeply entrenched problem, with profound implications for achieving SDG1 and enhancing urban health equity. These challenges originate from the historical uneven distribution of resources and the consequences of poorly planned land use and allocation strategies (Richmond et al. 2018; Watson 2009). The concentration of impoverished populations in distinct urban neighbourhoods such as Phomolong not only leads to their isolation but also hampers their ability to obtain essential services. Subsequently, a detrimental cycle of poverty and further widening the socioeconomic divide is established (Huchzermeyer 2019). The consequences of societal segregation extend beyond economic and social dimensions and impacts on urban health.

The case study of Phomolong demonstrated how poverty, both material (due to an over reliance on grants) and immaterial (due to poor building typologies, insufficient infrastructure, low education attainment and poor food security) contributed to a cycle of health inequalities and impeded poverty reduction (Corburn and Sverdlik 2019; Satterthwaite et al. 2019). Phomolong residents primarily construct their homes out of corrugated roofing sheets, which provide minimal protection against extreme weather conditions with the worsening climate change hazards in the country (van der Bank and Karsten 2020). According to Zerbo et al. (2020), such communities contend with continual health risks, which have a cumulative effect on their well-being and health. This illustrates the intertwined relationship between poverty and poor health, highlighting how one can be both a precursor and a result of the other.

The case of Phomolong also shows how poverty impairs a range of human qualities, and opportunities (Dean et al. 2017; Mani et al. 2013). The community's low educational attainment restricts their access to formal employment opportunities, trapping residents in low paying, informal jobs. As a result, there is an excessive reliance on frequently inadequate and unreliable government grants or subsidies.

To achieve SDG1 in South Africa, it is imperative to address disparities in marginalised communities such as Phomolong. The absence of essential infrastructure like clean water supply, sanitation, roads and electricity underscores the need for a strategic infrastructure plan by the Mangaung local authority. This entails giving precedence to providing fundamental services and ensuring lasting access to essential resources, ultimately enhancing the well-being and health of the community. Informal land tenure exacerbates vulnerability and limited development opportunities, underscoring the importance of legal recognition and mechanisms to ensure secure land rights. Consequently, improved spatial planning is essential for sustainable urban development and enhancing the quality of life for each resident, with a particular emphasis on health equity. Additionally, transportation and connectivity plays a role in shaping the socioeconomic landscape of communities (David et al. 2018; Gumede 2021). The lack of connectivity hinders upward mobility and exacerbates the challenges faced by those living in poverty (Evans et al. 2018). In poorly planned areas with inadequate connectivity, communities become isolated, making it difficult for residents to reach medical facilities (Dumedah et al. 2023). This isolation deepens health disparities, reduces healthcare–seeking behaviour among vulnerable populations, and exacerbates the health challenges they face (Matamanda et al. 2022).

Although there are commendable pro-poor strategies, there is still a need for concerted efforts to eradicate poverty and health disparities through equitable resource allocation in marginalized communities. With increasing climate change hazards, which are currently one of the main obstacles in achieving SDG1, there is pressing need to enhance resilience in housing and infrastructure to improve urban livelihoods. This proactive strategy lays the foundation for sustainable development. Furthermore, the state must consider community driven financing mechanisms to reduce informal settlement communities' over reliance on government grants. By empowering these communities to be responsible for their own development, South Africa can foster self-sufficiency and resilience, thereby enhancing long-term well-being.

2.6 Conclusion

In conclusion, Phomolong respondents' challenges with extreme poverty and urban health disparities present a nuanced picture of interconnected problems, reflecting a larger dilemma faced by many other poor communities in African cities. This study shed light on the complex variables that contribute to poverty, from unequal resource distribution and poor living conditions to poor transportation networks and environmental degradation. Beyond financial constraints, the repercussions could affect residents' physical and mental health, especially in impoverished areas. But despite its complexity, there remains room for transformative change. Improved basic infrastructure, and climate change resilient housing options as well as educational and employment opportunities are urgently needed to break the cycle of poverty and urban health disparities. The exploration of community driven financing mechanisms offers a possible path to long-term sustainability and reduced dependency on government grants. To achieve a sustainable, equitable and vibrant urban landscape for Bloemfontein, an integrated strategy is needed, which considers the intricate web of factors intertwined with spatial planning, health equity and inclusivity. Failure to implement these recommendations may undermine, not only the advancement of SDG1, but also other SDGs pertaining to infrastructure, urban development, health and education.

References

Aikins ER, Mclachlan JDT (2022) Africa is losing the battle against extreme poverty. Institute for Security Studies Retrieved 08 September from https://issafrica.org/iss-today/africa-is-losing-the-battle-against-extreme-poverty

Ayoo C (2022) Poverty reduction strategies in developing countries. In: de Salvo P, Piñeiro M (eds) Rural development – Education, sustainability, multifunctionality. IntechOpen

Azevedo JP (2020a) Learning poverty in the Time of COVID-19. https://openknowledge.worldbank.org/server/api/core/bitstreams/d977be42-3c37-5cb3-87ed-bd43eadbac1d/content

Azevedo JP (2020b) Learning poverty: measures and simulations. The World Bank

Bazilah A, Ibrahim AA, Yusof N, Jalani JJKH, Madini MHMM (2020) Policies on urbanization, poverty and hunger in South Africa: towards achieving SDG 1 and SDG 2. Brunei: Universiti Brunei Darussalam

Beegle K, Christiaensen L (2019) Accelerating poverty reduction in Africa [Accélérer la réduction de la pauvreté en Afrique]. World Bank Publications-Books

Bhorat H, Kanbur SR (2006) Poverty and policy in post-apartheid South Africa. HSRC Press

Bodo T (2019) Rapid urbanisation: theories, causes, consequences and coping strategies. Ann Geogr Stud 2(3):32–45

Bond P, Galvin M (2022) Durban flood exposes climate inaction. Green Left Weekly 15(1343)

Burns JK, Tomita A, Lund C (2017) Income inequality widens the existing income-related disparity in depression risk in post-apartheid South Africa: evidence from a nationally representative panel study. Health Place 45:10–16

Carr SC (2021) Setting 'poverty thresholds': whose experience counts? Sustain Sci 16(1):31–36

Castaneda R, Doan D, Newhouse DL, Nguyen M, Uematsu H, Azevedo JP (2016) Who are the poor in the developing world? World Bank Policy Research Working Paper (7844)

Chirisa I, Matamanda AR (2016) Addressing urban poverty in Africa in the post-2015 period. Perspectives for adequate and sustainable housing. J Sett Spat Plan 7(1)

CoGTA (2020) Mangaung Metropolitan Free State: profile and analysis district development model. C. G. a. T. Affairs. https://www.cogta.gov.za/ddm/wp-content/uploads/2020/08/DistrictProfile_MANGAUNG.01072020.pdf

Corburn J, Sverdlik A (2019) Informal settlements and human health. In: Nieuwenhuijsen M, Khreis H (eds) Integrating human health into urban and transport planning. Springer, pp 155–171. https://doi.org/10.1007/978-3-319-74983-9_9

Côté-Roy L, Moser S (2019) 'Does Africa not deserve shiny new cities?' The power of seductive rhetoric around new cities in Africa. Urban Stud 56(12):2391–2407

Cowling N (2023) Number of people living in extreme poverty in South Africa from 2016 to 2025. Statista Retrieved 14 September from https://www.statista.com/statistics/1263290/number-of-people-living-in-extreme-poverty-in-south-africa/#:~:text=As%20of%202022%2C%20around%2018.2,increase%20in%20the%20coming%20years

Cumming TL, Shackleton RT, Förster J, Dini J, Khan A, Gumula M, Kubiszewski I (2017) Achieving the national development agenda and the Sustainable Development Goals (SDGs) through investment in ecological infrastructure: a case study of South Africa. Ecosyst Serv 27:253–260

David A, Guilbert N, Hamaguchi N, Higashi Y, Hino H, Leibbrandt M, Shifa M (2018) Spatial poverty and inequality in South Africa: A municipality level analysis. Cape Town: SALDRU, UCT. (SALDRU Working Paper Number 221)

Dean EB, Schilbach F, Schofield H (2017) Poverty and cognitive function. In: The economics of poverty traps. University of Chicago Press, pp 57–118

Department of Human Settlements (2015) Towards a policy foundation for the development of human settlements legislation. D. o. H. Settlements. https://www.dhs.gov.za/sites/default/files/documents/TOWARDS%20POLICY%20FOUNDATION%20FOR%20THEHUMAN%20SETTLEMENTS%20LEGILATION%20-%2001%20NOVEMBER%202015%283%29%20%281%29.pdf

Dumedah G, Arthur J, Kokroko JS, Ampofo ST, Okyere PA, Tweneboah Kodua E (2023) Characterizing the geography of potential healthcare accessibility in informal settlements in the Greater Kumasi area of Ghana. Afr Geogr Rev 42(1):46–71

Evans J, O'Brien J, Ch Ng B (2018) Towards a geography of informal transport: mobility, infrastructure and urban sustainability from the back of a motorbike. Trans Inst Br Geogr 43(4):674–688

Finn BM, Cobbinah PB (2023) African urbanisation at the confluence of informality and climate change. Urban Stud 60(3):405–424

Fofana I, Chitiga-Mabugu M, Mabugu RE (2023) Is Africa on track to ending poverty by 2030? J Afr Econ 32(Supplement_2):ii87–ii98

Fonseca LM, Domingues JP, Dima AM (2020) Mapping the sustainable development goals relationships. Sustain For 12(8):3359

Förster T, Siegenthaler F (2018) Introduction: reimagining cities in Africa. Soc Dyn 44(3):395–404

Fourie W (2018) Aligning South Africa's National Development Plan with the 2030 Agenda's Sustainable Development Goals: guidelines from the policy coherence for development movement. Sustain Dev 26(6):765–771

Fourie H (2021) South Africa's progress towards its development objectives: 2020 Assessment Report. Stellenbosch: Bureau for Economic Research, Stellenbosch University

Francis E (2006) Poverty: causes, responses and consequences in rural South Africa. Chronic Poverty Research Centre Working Paper (60)

Francis D, Webster E (2019) Poverty and inequality in South Africa: critical reflections. Dev South Afr 36(6):788–802

Francis D, Valodia I, Webster E (2020) Politics, policy, and inequality in South Africa under COVID-19. Agrar South: J Polit Econ 9(3):342–355

Fransman T, Yu D (2019) Multidimensional poverty in South Africa in 2001-16. Dev South Afr 36(1):50–79. https://doi.org/10.1080/0376835X.2018.1469971

Gbadegesin J, Pienaar M, Marais L (2020) Housing, planning and urban health: historical and current perspectives from South Africa. Bull Geog Soc-Econ Ser 48:23–34

Graven MH (2014) Poverty, inequality and mathematics performance: the case of South Africa's post-apartheid context. ZDM 46:1039–1049

Gumede V (2021) Revisiting poverty, human development and inequality in democratic South Africa. Ind J Human Dev 15(2):183–199

Gunter A (2020) Pro-poor local economic development and the demand for economic growth: a case study on the lack of inclusive development in South Africa. In: Dilemmas of Regional and Local Development. Routledge, pp 229–242

Gunter A, Massey R (2020) South Africa's urban future: challenges and opportunities. In: Urban geography in South Africa: perspectives and theory. Springer International Publishing, pp 283–292

Gutura P (2014) Boosting Their Stomachsâ€™: the Role of Social Grants in South Africa. J Econ Behav Stud 6(2):105–119

Hakizimana J-M, Geyer H (2014) Socio-economic inequality in South Africa according to different disparity indices. 54th Congress of the European Regional Science Association: "Regional development & globalisation: Best practices", 26-29 August 2014, St. Petersburg, Russia. European Regional Science Association (ERSA), Louvain-la-Neuve

Huchzermeyer M (2009) The struggle for in situ upgrading of informal settlements: a reflection on cases in Gauteng. Dev South Afr 26(1):59–73

Huchzermeyer M (2019) Informal settlements and shantytowns as differential space. In: The Routledge handbook of Henri Lefebvre, the city and urban society. Routledge, pp 467–476

Huxham M, Anwar M, Nelson D (2019) Understanding the impact of a low carbon transition on South Africa. Climate Policy Initiative (CPI). https://climatepolicyinitiative.org/wpcontent/uploads/2019/03/CPI-Energy-Finance-Understanding-the-impact-of-a-low-carbon-transitionon-South-Africa-March-2019.pdf

International Monetary Fund (2023) World economic outlook update: near term resilience, persistent challenges. I. M. Fund. https://www.imf.org/en/Publications/WEO

Jansen A, Moses M, Mujuta S, Yu D (2015) Measurements and determinants of multifaceted poverty in South Africa. Dev South Afr 32(2):151–169

Jili NN, Masuku MM, Nzimakwe TI (2019) Enhancing pro-poor local economic development initiatives towards expediting sustainable development goals: King Cetshwayo District Municipality, KwaZulu-Natal. J Pub Admin 54(4–1):790–801

Juggernath S, Rampersad R, Reddy K (2011) Corporate responsibility for socio-economic transformation: a focus on broad-based black economic empowerment and its implementation in South Africa. Afr J Bus Manag 5(20):8224

Kim J, Hagen E, Muindi Z, Mbonglou G, Laituri M (2022) An examination of water, sanitation, and hygiene (WASH) accessibility and opportunity in urban informal settlements during the COVID-19 pandemic: evidence from Nairobi, Kenya. Sci Total Environ 823:153398

Kundu D, Pandey AK (2020) World urbanisation: trends and patterns. In: Developing national urban policies: ways forward to green and smart cities. Springer Nature, pp 13–49

Kwasi Fosu A (2018) Economic structure, growth, and evolution of inequality and poverty in Africa: an overview. J Afr Econ 27(1):1–9

Langford M, Cousins B, Dugard J, Madlingozi T (2013) Socio-economic rights in South Africa: symbols or substance? Cambridge University Press

Lavell A, McFarlane C, Moore HL, Woodcraft S, Yap C (2023) Pathways to urban equality through the sustainable development goals: modes of extreme poverty, resilience, and prosperity. Int J Urban Sustain Develop 15(1):215–229

Li X, Stringer LC, Dallimer M (2022) The impacts of urbanisation and climate change on the urban thermal environment in Africa. Climate 10(11):164

Mackett O (2020) Social grants as a tool for poverty reduction in South Africa? A longitudinal analysis using the NIDS Survey. Afr Stud Quart 19(1)

Mangaung Metropolitan Municipality (2021) Mangaung at the heart of it all: District/ Metro First Generation One Plan 2020–21. M. M. Municipality. http://www.mangaung.co.za/wp-content/uploads/2021/06/ITEM-11-2.District-Metro-First-Generation-One-Plan.pdf

Mani A, Mullainathan S, Shafir E, Zhao J (2013) Poverty impedes cognitive function. Science 341(6149):976–980

Manji K, Perera S, Hanefeld J, Vearey J, Olivier J, Gilson L, Walls H (2023) An analysis of migration and implications for health in government policy of South Africa. Int J Equity Health 22(1):1–16

Marais L (2021) Bloemfontein: three decades of urban change. In: South African urban change three decades after apartheid: homes still apart? Springer, pp 117–127

Marutlulle NK (2021) A critical analysis of housing inadequacy in South Africa and its ramifications. Afr Public Servi Deliv Perform Rev 9(1):16

Mashaba N, Saurombe MD (2023) Evaluating the effectiveness of a social grant regulating framework in South Africa. Acta Commer-Ind Res J Manag Sci 23(1):1011

Masilela T, Mthiyane S (2014) From the RDP to the National Development Plan: the mirage of a superministry. In: Meyiwa T, Nkondo M, Chitiga-Mabugu M, Sithole M, Nyamnjoh F (eds) State of the Nation 2014: South Africa 1994–2014: a twenty-year review. HSRC Press, pp 59–78

Matamanda AR (2024) Genesis of Hopley Farm informal settlement in independent Harare, Zimbabwe. Afr Geogr Rev 43(1):144–157

Matamanda AR, Dunn M, Nel V (2022) Broken bridges over troubled waters: COVID-19 and the urban poor residing in Dinaweng informal settlement, Bloemfontein, South Africa. S Afr Geogr J 104(3):309–327

Mazeka B, Sutherland C, Buthelezi S, Khumalo D (2019) Community-based mapping methodology for climate change adaptation: a case study of Quarry Road West informal settlement, Durban, South Africa. In: Cobbinah PB, Addaney M (eds) The geography of climate change adaptation in Urban Africa. Springer International Publishing, pp 57–88

McLean K (2009) Constitutional deference, courts and socio-economic rights in South Africa. Pretoria University Law Press (PULP)

Mphambukeli TN (2019) Social justice in planning: access to adequate drinking water and contested belonging in emerging communities of Mangaung in post-apartheid South Africa. African Renaissance 16(2):51–69. https://doi.org/10.31920/2516-5305/2019/SIn2a3

Mphambukeli TN, Nel V (2018) Migration, marginalisation and oppression in Mangaung, South Africa. In: Crisis, identity and migration in post-colonial Southern Africa. Springer International Publishing, pp 141–163

Mubecua MA, David OJ (2019) So far so good? Tracking the poverty eradication goal of SDGs in Kenya, Nigeria, and South Africa. J Public Aff 19(4):e1964

National Planning Commission (2012) National development Plan 2030: our future—make it work. https://www.gov.za/sites/default/files/gcis_document/201409/ndp-2030-our-future-make-it-workr.pdf

Ndasana M, Vallabh D, Mxunyelwa S (2022) Understanding the Service delivery protests in South Africa: a case study of Duncan Village. Afr Public Serv Deliv Perform Rev 10(1):7

Nel V (2011) Land-use management system as a tool towards achieving low-carbon cities in South Africa. Stads-en Streeksbeplanning= Town Reg Plann 2011(58):1–5

Nel V (2016) A better zoning system for South Africa? Land Use Policy 55:257–264

Ngepah N, Makgalemele T, Saba CS (2023) The relationship between education and vulnerability to poverty in South Africa. Econ Chang Restruct 56(1):633–656

Nhamo G, Dube K, Chikodzi D (2020) Sustainable development goals: concept and challenges of global development goal setting. In: Handbook of global health. Springer, pp 1–40

Nzimakwe TI (2020) Urbanisation and future smart cities: challenges of water and sanitation services. In: Reflections on African cities in transition: selected continental experiences. Springer Nature, pp 231–246

Oksiutycz A, Azionya CM (2022) Informal settlements: a manifestation of internal and cross-border migration. In: Migration in Southern Africa: IMISCOE regional reader. Springer International Publishing Cham, pp 109–124

Pieterse M (2004) Coming to terms with judicial enforcement of socio-economic rights. S Afr J Hum Rights 20(3):383–417

Pradhan P, Costa L, Rybski D, Lucht W, Kropp JP (2017) A systematic study of sustainable development goal (SDG) interactions. Earth's Future 5(11):1169–1179

Qizilbash M (1998) Poverty: concept and measurement. S. D. P. I. (SDPI). https://sdpi.org/sdpiweb/publications/files/R12-Poverty%20Concept%20and%20Measurement.pdf

Ramphoma S (2014) Understanding poverty: causes, effects and characteristics. Int: Inter J 13(2):59–72

Republic of South Africa (2020) Social Assistance Amendment Act, 2020. Government Gazette, Vol 666(44035)

Richmond A, Myers I, Namuli H (2018) Urban informality and vulnerability: a case study in Kampala, Uganda. Urban Sci 2(1):22

Satterthwaite D, Sverdlik A, Brown D (2019) Revealing and responding to multiple health risks in informal settlements in sub-Saharan African cities. J Urban Health 96:112–122

Schiel R, Leibbrandt M, Lam D (2016) Assessing the impact of social grants on inequality: a South African case study. In: Contemporary issues in development economics. Springer, pp 112–135

Schmidt U (2010) Poverty: a challenge to human dignity? Human dignity: a paradoxical concept. Diaconia 1(1):7–31

Stats SA (2021:4) National Poverty Lines 2021. D. o. S. S. Africa. https://www.statssa.gov.za/publications/P03101/P031012021.pdf

Triegaardt JD (2006) Poverty and inequality in South Africa: policy considerations in an emerging democracy. Annual Association of South African Social Work Education Institutions (ASASWEI) conference organised by University of Venda Department of Social Work

Turok I, Borel-Saladin J (2018) The theory and reality of urban slums: pathways-out-of-poverty or cul-de-sacs? Urban Stud 55(4):767–789

United Nations (2022) The sustainable development goals report 2022. D. o. E. a. S. A. (DESA)

van der Bank M, Karsten J (2020) Climate change and South Africa: a critical analysis of the earthlife Africa Johannesburg and another v Minister of Energy and Others 65662/16 (2017) case and the drive for concrete climate practices. Air Soil Water Res 13:1178622119885372

Verster P (2018) Transcendence in the context of abject poverty in informal settlements in Mangaung, South Africa. J Christian Scholar= Tydskrif vir Christelike Wetenskap 54(3–4):167–180

Watson V (2009) 'The planned city sweeps the poor away…': urban planning and 21st century urbanisation. Prog Plan 72(3):151–193

World Bank (2019) Ending learning poverty: what will it take?. W. Bank. http://hdl.handle.net/10986/32553

World Bank (2020a) Poverty and equity brief: Sub-Saharan Africa, South Africa. W. B. Group. https://databankfiles.worldbank.org/public/ddpext_download/poverty/33EF03BB-9722-4AE2-ABC7-AA2972D68AFE/Global_POVEQ_ZAF.pdf

World Bank (2020b) Poverty and shared prosperity 2020: reversals of fortune. W. Bank. https://openknowledge.worldbank.org/server/api/core/bitstreams/611fc6f2-140b-551e-9371-468eec64c552/content

World Population Review (2023) Bloemfontein population 2023. World Population Review. https://worldpopulationreview.com/world-cities/bloemfontein-population

Xulu NG, Chikoore H, Bopape M-JM, Nethengwe NS (2020) Climatology of the mascarene high and its influence on weather and climate over Southern Africa. Climate 8(7):86

Zarenda H (2013) South Africa's National Development Plan and its implications for regional development. Trade Law Centre, Stellenbosch, 8

Zerbo A, Delgado RC, González PA (2020) Vulnerability and everyday health risks of urban informal settlements in Sub-Saharan Africa. Glob Health J 4(2):46–50

No Hunger SDG and Urban Health in Mangaung

3

Maureen Maisiri, Moleboheng Pherane, Palesa B. Ayantokun, Leemisa S. Matooane, Mischka Dunn, and Abraham R. Matamanda

3.1 Introduction

Food insecurity is an increasing challenge that is currently being experienced across Africa. Hunger is a growing concern, which tends to affect the poor (Bellian 2019). We define hunger as the uneasy or painful sensation caused by a lack of food (Kendall et al. 1996). Persistent hunger causes people to starve which leads to malnutrition (Sasson 2012). Banovcinova et al. (2014) affirm that many families live below the poverty line, with some being at risk of poverty as most of these poorer households spend a greater proportion of their income on food (Drimie and McLachlan 2013; Sasson 2012). SDG 2 is aimed at ending hunger and improving food security (Gil et al. 2019; Harrington 2016; Shaibu 2021). Food security refers to the process in which people have access to sufficient food that is safe and nutritious to allow for a healthy lifestyle, thereby integrating the 3 pillars of sustainability (environment, social, and economic) (Fiddler 2012; Lifland 2012; Shaibu 2021). Food security ensures consistency in the availability of food. On the other hand, food insecurity refers to the point at which people are hindered from living an active and healthy life due to a lack of consistency in getting adequate food (Ryan et al. 2022; Fiddler 2012). Food insecurity causes people to be hungry which in effect impacts the health and quality of life (Kendall et al. 1996; Sasson 2012).

Through SDG #2, global leaders have committed to ensuring that there is a reduction in hunger by 2023, thereby leading to long-term food security, better nutrition and sustainable systems of production, distribution, and consumption of food resources (Disley 2013). Yet global statistics show that food insecurity is increasingly on the rise, especially in Africa, Latin America and Asia. While the FAO et al. (2023) report indicates that the prevalence of severe food insecurity at the global level declined slightly between 2021 and 2022 from 11.7% to 11.3% (approximately 27 million people fewer), there are discrepancies between regions which reveal the severity of the situation in other regions, especially for Africa. For example, hunger is more prevalent in Africa where close to 20% of people were identified to be facing hunger in 2023 while projections show that close to 600 million people will be chronically undernourished in 2030, making it difficult to achieve the sustainable development goals target to

M. Maisiri
Disaster Management Training and Education Centre, University of the Free State, Bloemfontein, South Africa

M. Pherane · P. B. Ayantokun · M. Dunn
Department of Geography, University of the Free State, Bloemfontein, South Africa

L. S. Matooane · A. R. Matamanda (✉)
Department of Urban & Regional Planning, University of the Free State, Bloemfontein, South Africa

A. R. Matamanda, V. Nel (eds.), *Sustainable Development Goals and Urban Health*, Sustainable Development Goals Series, https://doi.org/10.1007/978-3-031-68734-1_3

eradicate hunger. While food insecurity is projected to increase in Africa by 2030, specifically sub-Saharan Africa (SSA) is rapidly urbanizing. It is estimated that the population of sub-Saharan Africa SSA will quadruple by 2050, reaching 1.3 billion (UN DESA 2022).

Food is a basic need for everyone (Bellian 2019). Everyone needs to have access to food. Fiddler (2012) describes food as a need that historically was met through connection to the land. While food insecurity is on the rise, there are growing and emerging challenges and impacts that directly relate to urban health. Food is a critical pillar of urban life. The availability of adequate food is fundamental in achieving sustainable development because food plays a critical role in nourishing human bodies and providing energy to undertake all forms of socio-economic activities required to sustain cities. The relationship between adequate food and urban health is thus recognized (Haysom and Battersby 2023).

Food is a critical issue that is directly correlated with urban health. The lack of adequate food is characterized by several factors that compromise the health and well-being of urbanites, especially among the vulnerable demographic groups including children and the elderly who require specialized diets and nutrition for their well-being (Coleman-Jensen et al. 2022; Toossi and Jones 2023).

The phenomenon of hunger in a developing country such as South Africa should be minimal. As seen from a National perspective, South Africa is considered a food secure country (Tambe et al. 2023). However, poverty, the rising unemployment rates, socio-economic inequality and most the COVID-19 pandemic have negatively impacted the country and increased the number of people who suffer from malnutrition and hunger. This is shown in in 21% (3.72 million) of the South African population either have inadequate or severally inadequate access to food (Stats SA 2021). This is occurring despite policy-makers in South Africa setting expansive mandates that seek to ensure that their citizens have adequate access to food (Nkrumah 2020). Despite this very specific goal, South Africa has seen a sharp increase in the proportion of people suffering from hunger and undernourishment since 2011. The percentage of people suffering from hunger in the country went from 3.5% in 2011 to 5.7% in 2018 and likewise undernourishment increased from 1.8 million people in 2011 to 3.3 million in 2018 (Fan 2019). These figures are quite alarming and bring about a desire to probe into what are the factors contributing to it.

This chapter followed a phenomenological approach and participation in this study was voluntary. The participant's confidentiality and anonymity were ensured through the use of pseudonyms. Data for this study was collected using a qualitative approach and triangulation was used to gain a more detailed and extensive understanding of the level of food security and survival strategies of people regarding hunger. We did this by immensely engaging with the participant's observations. Through these observations in four different communities, we managed to capture how people practice urban agriculture as a means to combat poverty and a way of having access to fresh and nutritious food. We also undertook in-depth interviews with participants to understand their subjective experiences. We did this by assessing the level of urban health in Magaung with approximately 200 participants taking 50 respondents from each of the four areas. These became sources of information on the state of hunger and urban health in Mangaung.

3.2 Conceptual and Theoretical Approach

Lifestyle and poor choices are the leading causes of a lack of good nutrition (Fiddler 2012). Target 2.1 focuses on ensuring that universal access to safe and nutritious food is achieved. People living in urban areas are highly exposed to the unfavourable conditions of the cities that affect their health. People's nutritious diets get affected once they get comfortable with the affordability of fast food. Healthy food in urban areas is expensive, causing the majority not to have access due to poverty (Fiddler 2012; Bellian 2019; Sasson 2012; Wright et al. 2016). As a result, households are forced to live off unhealthy choices, as they are forced to live within a certain budget for survival (Banovcinova et al. 2014; Bellian 2019). Children exposed to such conditions end up growing up lacking necessary vitamins and good nutrition, intensifying undernourishment (indicator 2.1.1). However, Fiddler (2012) asserts that access to healthy and inexpensive food would lead to healthier healthy living.

The lack of food supply, combined with the expensive prices makes it difficult to achieve indicator 2.1.2, focused on the prevalence of food insecurity. Households not having enough food restricts the process of ending all forms of malnutrition (Target 2.2), as they end up not eating enough nutritious food, deteriorating their health. Children become vulnerable to abnormal growth, malnutrition, developmental delays, and low birth weight (Fiddler 2012; Shaibu 2021), thereby increasing the prevalence of childhood stunting due to recurrent malnutrition (indicator 2.2.1). Children also become susceptible to different illnesses and suffer from micronutrient deficiency (Shaibu 2021). This increases evidence of childhood malnutrition (indicator 2.2.2).

The structure of urban areas consisting predominantly of pavements and tar roads, together with the standards of living, leads to less effort being put into agricultural practices. Residents depend on the supermarkets for food supplies. Due to these circumstances, urban areas fail to work towards sustainable food production and resilient agricultural practices (Target 2.4), achieving sustainable food production (indicator 2.4.1) far within reach. With the biggest issue being the lack of agricultural development, Sasson (2012) argues that consistent food insecurity is caused by inadequate food production (Lifland 2012). Therefore, the fewer people promote agricultural practices, the harder it is to have adequate food production leading to hunger being persistent.

The food desert theory argues that less affluent people are subjected to eating poor diets due unaffordability and inaccessibility of fresh, healthy food in the poor neighbourhoods they live in (Wright et al. 2016). Food deserts focus on low income as they have a huge impact on families and children in lower-income areas (Bellian 2019; Wright et al. 2016). Wright et al. (2016) define food deserts as areas without ready access to fresh, healthy, and affordable food (Wright et al. 2016). The theory emphasizes that due to people experiencing a lack of food, they also lack access to variety and proper nutrition.

However, the theory has been criticised for putting too much emphasis on poor communities. Bellian (2019) reveals that the correlation between poverty and food insecurity is not the cause of food deserts. This is because some affluent people also find themselves eating food lacking nutrition, and some people living within the same neighbourhood may be eating completely different diets. Despite the criticism, this food deserts theory provides a transparent framework of how the poor neighbourhood in urban areas are affected in terms of food and health. With their concern highly concentrated on health, all measures to combat urban health issues are focused on an evolving society, to benefit everyone (Bellian 2019). Therefore, this theory has a huge impact on the quality of nutrition and overall health of inhabitants in urban areas.

3.3 Context of SDG 2 in South Africa

The discourse around Sustainable Development Goals (SDG's) has emerged as a powerful platform for addressing various challenges that affect modern-day society, whether in the rural or urban context. A worldwide agenda towards 2030 was provided for by the 17 Sustainable Development Goals that were endorsed by world leaders in 2015 (Veldhuizen et al. 2020). Of the 17 SDGs, SDG-2 aims to "End hunger, achieve food security and improve nutrition and promote sustainable agriculture" (FAO 2016). Furthermore, extant literature suggests that SDG-2 is essential to the achievement of the overall SDG agenda since it is inextricably linked to the economy, the environment, and society (FAO 2016; Gill et al. 2019). However, as it stands, SDG-2 is not expected to be accomplished by 2030. Why? Because about 828 million people globally are still experiencing severe hunger and or undernourishment (WHO 2022) and one-third of the global population is affected by some form of micro-nutrient deficiency (Han et al. 2022).

While SDG-2 is aimed at eradicating hunger and its consequences globally, achieving its targets and indicators will require several trade-offs and synergies with other SDGs at different temporal and spatial scales. This will therefore highlight how integrated the SDG agenda is (Gil et al. 2019). This goes on to show the SDGs were intentionally formulated to be broadly applicable in various country set-ups. Hence each SDG is supported by indicators and targets. The targets were defined as time-specific outcomes that contribute toward the goal of zero hunger while the indicators were to be specifically used to track a target's advancement, usually based on data that is already accessible (Roser 2023). SDG-2 is therefore defined by eight targets and fourteen indicators to be met by 2030 (see Fig. 3.1).

3.4 SDG-2: A Critical Component of the Broader Sustainable Development Agenda

One key understanding that underpins the SDGs is that each of them has a synergistic relationship with one another (Atukunda et al. 2021). Thus, achieving the goal of one SDG has an impact on achieving the goal of others. For instance, SDG-2 has an impact on SDG-1 in that, reducing poverty is intrinsically related to ensuring that everyone has sufficient access to healthy food. Hence, addressing undernutrition can not be separated from tackling poverty (Mollier et al. 2017; Christiaensen and Martin 2018). There is a synergy between SDG-2 and SDG-3 "Good health and well-being" because one of the most essential factors influencing health and well-being is consuming a balanced diet. However, globally, malnutrition continues to exert more strain on the healthcare systems thereby reducing the progress towards meeting the goals of both these SDGs (Adeyeye et al. 2023). According to WHO (2022), about 45% of deaths of children under the age of five years are directly related to

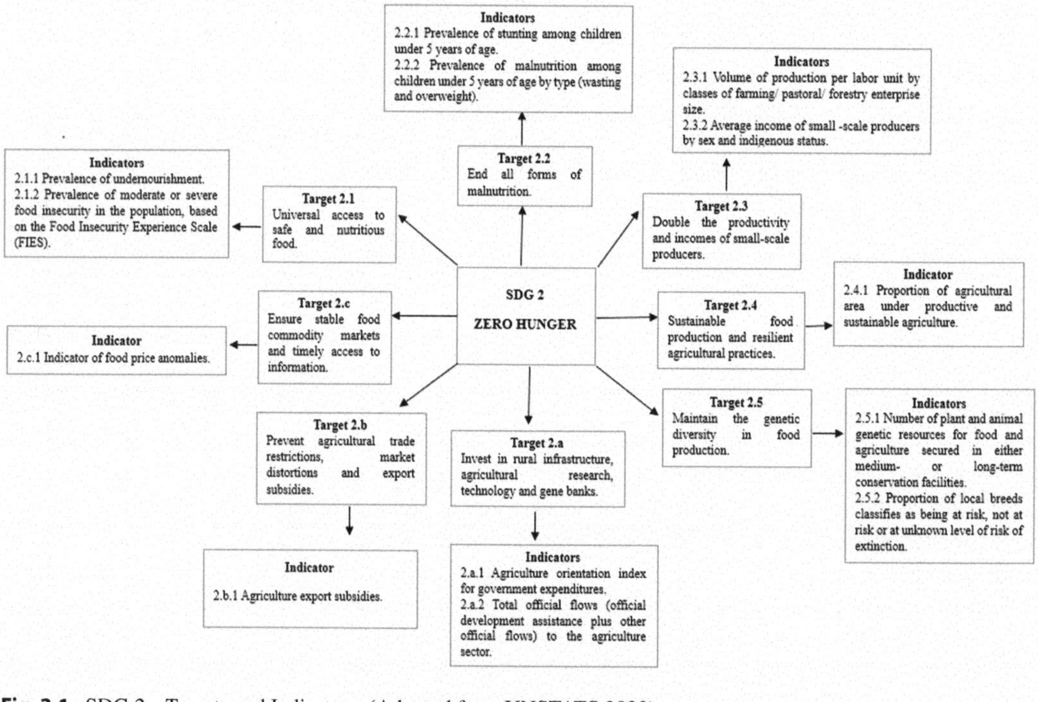

Fig. 3.1 SDG 2 – Targets and Indicators. (Adopted from UNSTATS 2023)

malnutrition. In other words, malnutrition poses severe threats to one's health and well-being. Agriculture therefore has a direct impact on mental and physical health since it is responsible for the provision of food for household consumption (Mollier et al. 2017; Adeyeye et al. 2023).

Based on the aforementioned, an enhanced food and nutrition status is an essential driver for good health and well-being. Another essential relationship exists between SDG-2 and SDG-4 "Quality education" because, since lack of adequate food results in malnutrition, a compromised nutritional status in young children impairs cognitive capacity and may result in potentially irreversible effects (Latham and Cobos 1971; Osei and Lambon-Quayefio 2022). Therefore, malnutrition hinders education by exacerbating the detrimental consequences of numerous other aspects of poverty and is linked to reduced concentration, reduced school attendance, and early school dropout rates. For that reason, a child's ability to learn and their nutrition are mutually supportive, just as health outcome and nutritional status are closely related.

Agriculture is at the center of food provision and it is also one of the main contributors to the GDP of many economies the world over. As a result, many of the most vulnerable and impoverished people rely on agriculture for their livelihoods thus promoting economic development in poor communities. Therefore, SDG-8 "Decent work and economic growth" has a relationship with SDG-2 because improving urban agriculture and employment opportunities result in sustainable economic growth which in turn allows people to have better health outcomes and reduce food insecurity in urban households (Kawachi and Kennedy 1999; Mollier et al. 2017). In a broader sense, population health patterns may be influenced by economic factors (Kawachi and Kennedy 1999).

Evidence shows that disparities in urban health reflect deeper social inequalities (Todd 1996). Therefore, if urban food insecurity is reduced by making food more accessible and more affordable to urban dwellers the goals of SDG-2 can be achieved. This shows that achieving the targets of SDG-2 can influence achieving the equality targets under SDG-10 "Reduced inequalities." Considering the relationship between SDG-2 and other SDG it can be said that SDG-2 is an integral part of the 2030 agenda in promoting urban health globally.

3.4.1 Towards Zero Hunger: Priorities and Achievements in South Africa

Based on the tenets of SDG-2, the goal is to eradicate hunger by 2030. However, since 2015, there has been a significant spike in global food insecurity, hunger, malnutrition, and or undernutrition. This spike has been made worse by some reasons, such as the COVID-19 pandemic, armed conflict in Ukraine, the Middle East, Sudan, and the Sahel, climate change, and growing inequalities among nations (Chen et al. 2023). Complementary to this, statistics show that in 2019, 8.9% of the global population was experiencing severe hunger and undernourishment and in 2022 the number had sharply risen to 9.2% (Valin et al. 2021). Additionally, 2.4 billion people worldwide experienced mild to severe food insecurity. Compared to 2019, this number increased by a staggering 391 million individuals (Valin et al. 2021). This categorization indicates that people have no access to enough food thus posing a threat to their overall health outcomes (Laborde et al. 2021; Valin et al. 2021).

Projections show that more than 600 million people will be facing extreme hunger globally in 2030, thereby underscoring the substantial difficulty of achieving the goal of zero hunger (WHO 2021). Severe hunger and malnourishment continue to obstruct sustainable development while setting a trap that is difficult for people to escape from. A population that is hungry and malnourished is less productive, more susceptible to sickness, and frequently unable to increase their income and standard of living.

While SDGs are not legally binding, governments across the world have realized the importance of sustainable development for humanity to thrive (Yang et al. 2020). Therefore, there have been continuous efforts to monitor and evaluate the progress of each country in meeting the goals of each SDG (Sachs et al. 2019). However, due to the varying dynamics of each country, various SDGs are taking priority over others. For example, the SDGs on health and well-being, equality, industrialization, employment and food security are the most prioritized in Africa (D'Alessandro and Zulu 2018; Yang et al. 2020). Despite this, the central themes around SDGs remain as the eradication of poverty and reinforcement of equity (Sachs et al. 2019). Moreover, Yang et al. (2020) emphasize that SDG-2, SDG-1 and SDG-6 have the highest priorities at the global and regional level.

3.4.2 Current Status of SDG-2 in South Africa

The effective implementation of SDGs largely depends on the goodwill of national governments. The South African government as well as the private sector, non-profit organizations (NPOs), and academic institutions have been effectively raising awareness towards the realization of the SDGs. South Africa is prioritizing all the 17 SDGs and progress has been made in the monitoring of SDGs and data provision thereof. At the global level, South Africa is ranked 110 out of the 166 countries that have committed to address the SDGs and at the country level, it has shown limited progress towards the achievement of all the SDGs with a score of 64% (Fig. 3.2). In terms of data coverage, as of 2023, the country was able to achieve 72.3% coverage (175 of 242) of the SDG indicators (STATS SA 2023). Despite the impressive coverage in some areas, there are still some major gaps in the geographic coverage, timelines and data disaggregation.

For SDG-2 South Africa prioritized the following indicators under SDG-2; prevalence of undernourishment, prevalence of stunting in children under 5 years of age, prevalence of wasting in children under 5 years of age, prevalence of obesity (BMI >30), human trophic level, cereal yield, sustainable nitrogen management index, export of hazardous pesticides. The data however, shows that South Africa is not progressing well enough to meet the overall goal of SDG-2 by 2030. Indications show that it is in the danger zone (red threshold) (Fig. 3.3) (Otekunrin 2021).

Despite lagging behind in meeting the overall goal of zero hunger by 2030, South Africa has achieved the green threshold for two indicators relating to health namely; prevalence of undernourishment and prevalence of wasting in children under 5 years of age (Table 3.1).

3.4.3 Achieving Zero Hunger: Implications on Urban Health

According to Chisasa and Khumalo (2023), about 67.35% of South Africa's population was living in urban areas by 2020 and the number is predicted to have risen to 71.3% by 2030. In addition, eight out of ten people will be residing in urban areas by 2050, thereby increasing the need for more food and basic infrastructure. Furthermore, there is an influx of young people who are migrating from rural areas to urban areas to seek better employment opportunities (Bhanye 2023). Hlongwane and Vyas-Doorgapersad (2023) support the notion that there are challenges that come with an increasing urban population. For example, in the case of South Africa low-income wages in urban areas have resulted in urban dwellers compromising on some of the requirements for a decent living such as adequate food (Mkhize et al. 2023). About 25.6% of the total South African population lives in slums and lack access to decent lives that include access to healthy and nutritious food and diets (Saint et al. 2023).

If the adult population in urban areas is living in poverty it not only impacts adults but also affects the well-being of children and the elderly because they will be limited to accessing nutritious food daily. Limited access to food results in undernourishment which in turn results in stunting. Children who experience stunted growth are likely to suffer from cognitive, intellectual, and developmental challenges, not only in their childhood but also in their adult life (Goudet et al. 2017). Also, undernourishment not only affects human health but also the expansion of the

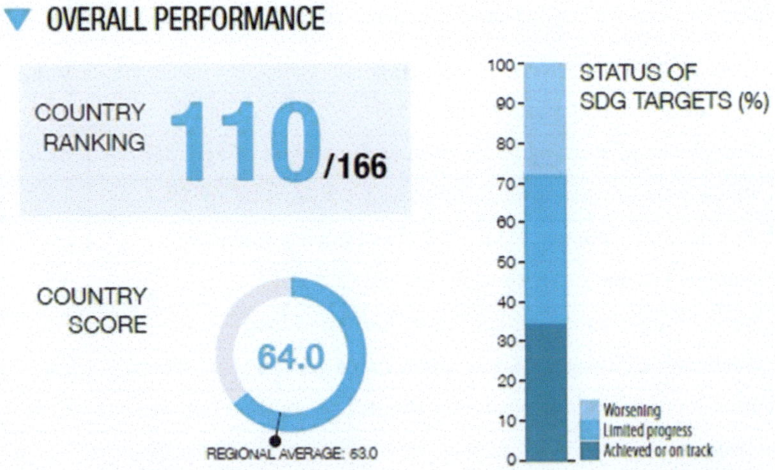

Fig. 3.2 Status of South Africa in meeting SDG targets. (Source: Sustainable Development Report 2023)

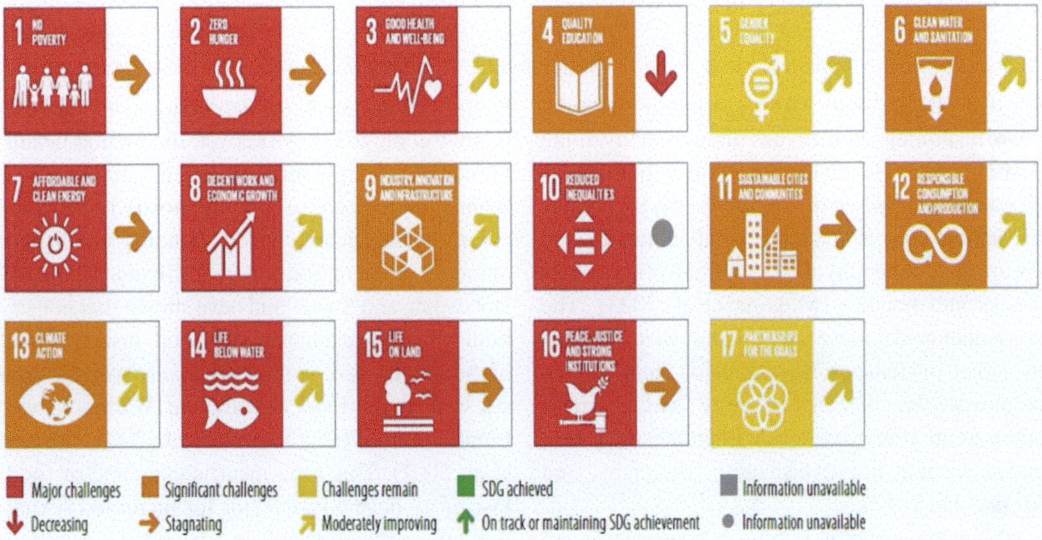

Fig. 3.3 Status of SDG-2 (Zero hunger) in South Africa. (Source: Sustainable Development Report 2023)

country's economy because more of the financial resources will be channeled to address the healthcare challenges related to this (Smith and Haddad 2000; Alderman et al. 2003; Goudet et al. 2017).

A household's ability to afford food is greatly influenced by employment. Results from household surveys conducted in 2021 showed that 19% of households in South Africa without an employed member experienced hunger (STATSSA 2023). Given that the unemployment rate is on the rise and is currently at 31.9%, this will have or already has detrimental effects on achieving goal 2 of zero hunger by 2030 in South Africa (STATSSA 2023). Furthermore, the biggest metros in South Africa (The City of Cape Town and the City of Johannesburg) are linked to a high percentage of households reporting very poor and insufficient access to food. What does this mean for urban health?

A rise in nutrient deficiency diseases in both children and adults because people will not prioritise healthy components of the diet but will focus on staple-rich diets (Battersby and McLachlan 2013). On the contrary, while hunger is experienced in about 19% of urban households in South Africa, obesity has been prevalent

Table 3.1 Status of SDG-2 Indicators in South Africa

SDG-2 – ZERO HUNGER: Indicators	Value	Year	Rating	Trend
Prevalence of undernourishment (%)	6.9	2020	● (green)	→ (stagnating)
Prevalence of stunting in children under 5 years of age (%)	22.8	2022	● (red)	↓ (decreasing)
Prevalence of wasting in children under 5 years of age (%)	3.8	2017	● (green)	● (information unavailable)
Prevalence of obesity, BMI ≥ 30 (% of adult population)	28.3	2016	● (red)	↓ (decreasing)
Human Trophic Level (best 2–3 worst) 2.3 2017	2.3	2017	● (yellow)	→ (stagnating)
Cereal yield (tonnes per hectare of harvested land)	5.1	2021	● (green)	↑ (on track)
Sustainable Nitrogen Management Index (best 0–1.41 worst)	0.8	2018	● (red)	↗ (moderately improving)
Exports of hazardous pesticides (tonnes per million population)	95.6	2020	● (red)	● (information unavailable)

■ Major challenges ■ Significant challenges ■ Challenges remain ■ SDG achieved ■ Information unavailable
↓ Decreasing → Stagnating ↗ Moderately improving ↑ On track or maintaining SDG achievement ● Information unavailable

Source: Sustainable Development Report (2023)

(BMI > 30) in South Africa's urban population. Obesity among women has increased over the past 10 years. The diet in informal settlements is characterized by low dietary diversity, high-calorie intake, low fruit and vegetable intake, and cheap and unhealthy food alternatives such as snacks and noodles (Mokone et al. 2023). The consequences of these are apparent as it will lead to a rise in diabetes, hypertension, and related cardiovascular diseases thereby adding more strain to an already congested healthcare system and lowering life expectancy (Battersby and McLachlan 2013).

While the general status of South Africa in terms of meeting the goals of SDG-2 is in the danger zone more than halfway towards 2030, the implications of hunger and food insecurity at the household level in urban areas on mental health are worrying. Gyasi et al. (2020) highlight that experiencing hunger and food insecurity leads to mental health problems. Even though mental health issues are still shunned in most African countries and also lacking government investment due to competing health and other developmental priorities we cannot neglect the impact of hunger and food insecurity on mental health (WHO 2022). In a study carried out in New Zealand, it was discovered that there is a strong effect of food insecurity and mental health complications such as depression and anxiety (Carter et al. 2011). Moreover, the mental health burden of food insecurity has been found to affect women more because they are mostly homemakers in the African context. Studies conducted among women in South Africa, Ghana, Thailand, Botswana, and Swaziland have shown that households that were chronically food insecure had higher numbers of women who were worried and stressed about food (Avotri and Walters 1999; Lemke et al. 2003; Piaseu et al. 2004; Weiser et al. 2007). This food insufficiency and or food insecurity pose a risk factor for increased sexual risk-taking which results in Sexually Transmitted Infections (STIs) among women which negatively affects their general health outcomes (Weiser et al. 2007).

It is undeniable that hunger and poverty are interconnected and they largely contribute to a population's health outcomes. South Africa is falling behind in achieving many of the targets of the United Nations Sustainable Development Goals and this exposes the population to vulnerabilities thereby affecting the developmental progress that a country can make. The failure of SDG-2 in South Africa will prolong food insecurity and create an unending cycle of poverty in mostly the urban population.

3.5 Results from Mangaung

3.5.1 Food Insecurity and Hunger

First, food security is a major problem that the MMM currently grapples with (MMM 2020). Food insecurity is evident from child hunger that has been reported across the metro in recent years with different organisations trying to combat the problem. For example, different private companies and NGOs have continuously contributed to the Rise Against Hunger initiative aimed at alleviating child hunger among children from underprivileged communities (News24 2019). However, in 2015, the Free State Project highlighted that food insecurity was prevalent in both formal and informal households revealing the increasing severity of the food insecurity problem. This food crisis and hunger in the MMM have persisted over the years largely due to food inflation, climate change-induced droughts leading to crop failure, and the COVID-19 pandemic. As of 2021, it was reported that 11,404 households in MMM had children who experienced hunger (News24 2019). In extreme cases, it has been highlighted that some children have died due to hunger and starvation (Oppelt 2024). While death is the extreme case, the burden of child hunger has multiple health challenges that include malnutrition, stunting, and impaired development (Alderman et al. 2003). Such confirm studies by Goudet et al. (2017) who have reported that children experiencing stunted growth are likely to suffer from cognitive, intellectual, and developmental challenges from childhood into adulthood.

Food insecurity and successful food production in Mangaung are threatened due to climate change-induced stressors including heatwaves, droughts, and flooding. According to Mbilemi (2015: 83), MMM faces climate change risk that causes increased food insecurity and breakdown of food production and supply systems due to increased temperatures, drought, flooding, rainfall variability, and other extreme weather events. Specifically, in the last decade, climate change-induced droughts have affected local farmers in the MMM with negative impacts on crop production. In 2018, it was reported that local farmers were struggling to make ends meet which also culminated in food insecurity (Mtembele 2018). The weather conditions of MMM characterized by extremes of temperature and rainfall also make it difficult for households to successfully engage in urban agriculture. In Phomolong, the failure of urban agriculture was noted as shown in Fig. 3.4. Such conditions may explain why few households in the MMM engage in urban agriculture.

Although one of the targets highlighted by SDG#2 to combat or address this was to double the agricultural productivity, this reality is not being realized in the Mangaung Municipality, according to the Census data from 2022, from the 231,921 surveyed households, only 46,172 households were involved in agricultural activities. This amounts to only 19% percentage of Mangaung Municipality households trying to address this crisis through agricultural produce. In a survey we conducted in Caleb Motsabi informal settlement in Bloemfontein it was revealed that water challenges were among the main stressors limiting vegetable production as shown in Fig. 3.4. The limited space for vegetable production was also identified by some residents as limiting their chances to engage in urban agriculture, thereby making it difficult for them to access fresh vegetables. Such issues show the challenges inherent in informal settlements and the vulnerabilities that are exacerbated by several other structural factors.

3.5.2 Energy and Food Preparation

Food preparation in poor neighburhoods was identified as a major challenge that compromised the quality and nutritional value of the food. Some residents from Phomolong informal settlement lamented that the lack of electricity and clean energy was an issue affecting the quality of their food. As shown in Fig. 3.5, there are some challenges with using firewood, which is not readily available in Phomolong and makes it difficult for families to prepare meals when it rains. One respondent indicated that they are forced to consume cold food as they fail to warm some leftovers thus compromising their health as

Fig. 3.4 Vegetable garden that is not doing well due to limited water and extreme temperatures

Fig. 3.5 Food being cooked outside

they risk some stomach problems from consuming such food. Paraffin stoves, as shown in Fig. 3.6 are also used by some households but one respondent stated that there are inherent health problems associated with using paraffin stoves indoors which affects their respiratory system. This is evidence of how the poor are affected also in their efforts of trying to prepare a meal ought to nourish their bodies.

Due to the challenge of energy for cooking, residents in the informal settlements (Phomolong and Caleb Motshabi) pointed out that they resort to other alternative foods such as cool drinks, snacks, and bread which are easier to prepare. However, these alternative foods are less nutritious and do not sustain a balanced diet leading to nutritional deficiency among the community members. A respondent commented that:

> It is especially hard for us people who live in parts that don't have electricity, we have to buy food like tin fish, some of us don't like it but you have to buy such foods just to have provision for the dark days. The rest of the family will eat it and while I will remain just like that. There are no fridges, we have to buy paraffin and it finishes quickly.

3.5.3 University Students Experiencing Hunger

An overlooked group that experiences food insecurity has been university students in the MMM. Considering the increasing cost of living and the limited funding that some university

Fig. 3.6 Paraffin stove used to prepare meals indoors

students have, food poverty is increasingly becoming a challenge among many students. A study by Van den Berg and Raubenheimer (2015) found that a significant number (60%) of students from their sample of students from the University of the Free State (UFS) experienced food insecurity with hunger. A critical issue noted in this study regarding university students was the unavailability of adequate food services on campus that students could rely on to have healthy meals as some were not confident with their cooking. This situation reveals some form of food desert inherent among university students. University students interviewed complained of mental health problems associated with the food insecurity they experienced. Some highlighted that "it's traumatizing when you have to be studying but you know you have nothing to eat. The mental stress is just too much to deal with." This perspective reveals the magnitude of the hunger issue among university students and how they remain marginalized and yet suffer in silence.

However, there have been several initiatives adopted by the UFS to address the food challenges among students. The following are selected examples:

- The No Student Hungry Campaign introduced at the UFS ensures that all students in need of food assistance are helped thus addressing the issues raised by Van den Berg and Raubenheimer (2015) and ensuring students are well-fed and nourished.
- Introduction of a food bursary that supports students experiencing financial difficulties and failing to meet their food needs.
- Establishment of the community garden which produces different vegetables that are donated to vulnerable students, ensuring they have access to healthy and nutritious vegetables.

3.5.4 COVID-19 Impact on Hunger

Second, the COVID-19 pandemic has further exacerbated food insecurities in MMM, especially among marginal communities. The situation has been dire for individuals and communities residing in informal settlements. During the COVID-19 pandemic, food systems were affected in multiple ways that eventually hindered the progress toward ending hunger and starvation in the MMM. The increase in food prices negatively affected the poor households who failed to access some basic commodities. In this instance, food inflation meant some households could not afford to sustain their food requirements. Some respondents remarked that,

> They are increasing food prices affected them badly because even when specials are made, you can't see the difference. They now only buy the important things. They can't afford to buy things like shoes. They are not able to do proper groceries, anyone, they have become very strict on how they eat. The priorities are now having pap and paraffin.

The above quote illustrates the challenges emanating from food price increases in 2020 during the peak of the COVID-19 pandemic. Another respondent highlighted that "*When the food prices increased, it was very stressful. To cope with this, we reduced the amount of food that we would buy. So, we were always running low on food.*" This quote illustrates the health impacts associated with escalating food prices which are the mental health challenges household heads suffered through anxiety and emotional stress in trying to make ends meet. Moreover, cutting down on certain food types, skipping meals, or eating less was adopted by some households, but this also has some health implications as individuals are not well-nourished by the food they eventually consume.

Subsequently, some households relied on social grants as their only sources of income, but these were meagre and could not cater to all their needs as explained by the respondent that.

> In terms of food, we struggled because the only income we had was that from the children's grant and that was not enough, so whenever we would buy food, it wouldn't last the whole month.

3.5.5 Sustainable Agriculture and Food Production

While food production has been affected by climate change, there have been some strides made to enhance food production at community levels. In this regard, there have been initiatives by some corporates and non-governmental organizations (NGOs) to support community development projects and agricultural production to ensure families have nutritious diets and meals aimed at meeting target 2 of SDGs #2. The MMM SDF (2020–2036) has provided for action plans that seek to enhance food production in the metropolitan area. Action 6.7 focuses on "*utilizing precision farming to minimize the impact of agriculture on natural resources*" (MMM 2020: 165). Specifically, "smart development" is envisaged in the metro through which controlled environment farming will be undertaken in Botshabelo and Thaba Nchu on the vacant factory shells located in the industrial areas. This is a great initiative aimed at enhancing food production and protecting the environment. Efforts have also been earmarked to link this industrial area to become part of the value chain in the proposed Estoire-N8 industrial area close to the Bloemfontein Fresh Produce Market (MMM 2020). Through the development of this value chain, it becomes critical to link farmers to markets thus allowing for food availability and thus limiting food scarcity and wastage as farmers at times fail to access markets.

Figure 3.7 shows some vegetables being grown in a greenhouse in Phomolong informal settlement. This initiative is aimed at enhancing food production and residents commended the project for enabling them to get fresh vegetables. Likewise, ForAfrika's Mangaung Food Security Programme has partnered with Woolworth Trust to enhance food security in Thaba Nchu through the establishment of community gardens. This initiative has been critical in addressing the need for SDG#2 target 2.1. The programme also supports small farmers through capacity development and ensures they are equipped with the requisite skills and knowledge to improve food production. Box 3.1 shows the success of the Mangaung Food Security Programme, highlighting the achievements and progress made towards SDG#2 in Thaba Nchu

Box 3.1

- Farmer turned a dumpsite into a viable food garden, ensuring food for his family and neighbours,
- Farmers selling their produce to local businesses and the communities they reside in,
- Farmer growing vegetables used in her soup kitchen feeding 150 people daily,
- Two farmers have been identified as mentors to the community,

431 community members benefit from the programme.

Fig. 3.7 Spinach being grown in a greenhouse in Pomolong

> It was hard to buy food during the lockdown because my husband was forced to stay at home and couldn't work and since I am not working it was very hard. We survived on the children's grant money.

3.5.6 Marketing Food Produce

The distribution of food markets and retailers is relatively good in MMM. It emerged that there are different food retailers in MMM. First, ther are retail shops and spaza shops that cater for the needs of the community. In informal settlements, community members decried the absence of retail shops which forced them to buy from spaza shops which did not always have adequate food. This situation reveals the food desert experienced by the communities who are then forced to consume alternative foods or have to forego some nutritious foods.

3.6 Discussion and Conclusion

Urbanisation in South Africa is steadily rising, with projections indicating a significant increase in urban population percentages by 2030 and 2050. This increase intensifies the demand for essential resources like food and infrastructure.

Young individuals are migrating from rural to urban areas in search of better job prospects, leading to demographic changes. However, this urbanization poses challenges, particularly regarding low-income wages, which force urban dwellers to compromise on necessities like adequate and nutritious food (Hlongwane and Vyas-Doorgapersad 2023). A considerable portion of the population resides in informal settlements, lacking nutritious food and decent living conditions. Poverty in urban areas not only affects adults but also has detrimental effects on children's well-being, resulting in stunted growth and long-term developmental issues (Dean et al. 2023; Hussain et al. 2023).

Undernourishment not only impacts health but also strains the country's economy, redirecting financial resources towards healthcare (Smith and Haddad 2000; Alderman et al. 2003; Goudet et al. 2017). Household surveys indicate a significant correlation between employment and food insecurity, with a rising unemployment rate threatening the achievement of zero-hungry goals by 2030 (STATSSA 2023). Major urban centers like Cape Town and Johannesburg face significant challenges with food access, posing risks to urban health. Unhealthy dietary practices contribute to rising obesity rates, alongside nutrient deficiency diseases, leading to increased instances of diabetes and cardiovascular issues (Battersby and McLachlan 2013).

Mental health challenges associated with food issues have also been highlighted showing the severity of hunger and food insecurity issue. This has been noted among the university students, but the situation is also prevalent among the poor residing in the informal settlements.

Initiatives implemented by various organizations in the MMM have made an effort to combat the zero-hunger initiative but with rising temperatures, climate change, and a lack of resources and infrastructure, the metropolitan has a long way to go in realizing significant success in addressing food insecurity. Failure to address hunger and poverty undermines progress towards Sustainable Developmental Goals, perpetuating cycles of vulnerability and poverty in urban populations. Overall, addressing these interconnected issues is crucial for improving health outcomes and achieving sustainable development in South Africa.

References

Adeyeye SAO, Ashaolu TJ, Bolaji OT, Abegunde TA, Omoyajowo AO (2023) Africa and the Nexus of poverty, malnutrition and diseases. Crit Rev Food Sci Nutr 63(5):641–656

Alderman H, Hoddinott J, Kinsey B (2003) Long-term consequences of early childhood malnutrition. FCND Discussion Paper No. 168. International Food Policy Research Institute, Washington, D.C.

Atukunda P, Eide WB, Kardel KR, Iversen PO, Westerberg AC (2021) Unlocking the potential for achievement of the UN Sustainable Development Goal 2–'Zero Hunger'–in Africa: targets, strategies, synergies and challenges. Food Nutr Res 65

Avotri JY, Walters V (1999) You just look at our work and see if you have any freedom on earth: "Ghanaian women's accounts of their work and their health". Soc Sci Med 48(9):1123–1133

Banovcinova A, Levicka J, Veres M (2014) The impact of poverty on the family system functioning. Procedia Soc Behav Sci 132:148–153

Battersby J, McLachlan M (2013) Urban food insecurity: a neglected public health challenge. S Afr Med J 103(10):716–717

Bellian R (2019) Food Deserts in Urban Areas. Downtown Rev 6(1):1–9

Bhanye J (2023) "Strategic, dual sense of place" among middle-aged migrants in the diaspora. Int J Migr Health Soc Care 19(3/4):376–390

Carter KN, Kruse K, Blakely T, Collings S (2011) The association of food security with psychological distress in New Zealand and any gender differences. Soc Sci Med 72(9):1463–1471

Chen X, Shuai C, Wu Y (2023) Global food stability and its socio-economic determinants towards sustainable development goal 2 (Zero Hunger). Sustain Dev 31(3):1768–1780

Chisasa J, Khumalo P (2023) Exploring the nexus between migration and urbanisation in the Southern Africa region. International Journal of Research in Business and Social. Science (2147–4478) 12(10):85–95

Christiaensen L, Martin W (2018) Agriculture, structural transformation and poverty reduction: eight new insights. World Dev 109:413–416

Coleman-Jensen A, Rabbitt MP, Gregory CA, Singh A (2022) Household food security in the United States in 2021, ERR-309. U.S. Department of Agriculture, Economic Research Service, Washington, D.C.

D'Alessandro C, Zulu LC (2018) From the Millennium Development Goals (MDGs) to the Sustainable

Development Goals (SDGs): Africa in the post-2015 development agenda. A geographical perspective. In: Africa in the Post-2015 Development Agenda. Routledge, pp 1–18

Dean G, Vitolins MZ, Skelton JA, Ip EH, Lucas CB, Brown CL (2023) The association of food insecurity with mental health in preschool-aged children and their parents. Pediatr Res 94(1):290–295

Disley P (2013) Sustainable development goals for people and planet. Nature 495:306–307

Drimie S, McLachlan M (2013) Food security in South Africa—First steps toward a transdisciplinary approach. Food Secur 5:217–226

Fan S (2019) Food policy in 2018–2019: growing urgency to address the SDGs. International Food Policy Research Institute (IFPRI), 2019 global food policy report, pp.6–15

FAO, F (2016) Food and agriculture: key to achieving the 2030 agenda for sustainable development. Job No. I5499. Food and Agriculture Organization of the United Nations, p.23

FAO, IFAD, UNICEF, WFP, WHO (2023) The state of food security and nutrition in the world (2023) Urbanization, agrifood systems transformation and healthy diets across the rural-urban continuum. FAO, Rome

Fiddler T (2012) Food security in a northern First Nations community: an exploratory study on food availability and accessibility. Int J Indig Health 8(2):5–14

Gil JDB, Reidsma P, Giller K, Todman L, Whitmore A, van Ittersum M (2019) Sustainable development goal 2: improved targets and indicators for agriculture and food security. Ambio 48(7):685–698

Gill JC, Mankelow J, Mills K (2019) The role of Earth and environmental science in addressing sustainable development priorities in Eastern Africa. Environ Develop 30:3–20

Goudet SM, Kimani-Murage EW, Wekesah F, Wanjohi M, Griffiths PL, Bogin B, Madise NJ (2017) How does poverty affect children's nutritional status in Nairobi slums? A qualitative study of the root causes of undernutrition. Public Health Nutr 20(4):608–619

Gyasi RM, Obeng B, Yeboah JY (2020) Impact of food insecurity with hunger on mental distress among community-dwelling older adults. PLoS One 15(3):e0229840

Han T, Dong J, Zhang J, Zhang C, Wang Y, Zhang Z, Xiang M (2022) Nutrient supplementation among pregnant women in China: an observational study. Public Health Nutr 25(6):1537–1542

Harrington LMB (2016) Sustainability theory and conceptual considerations: a review of key ideas for sustainability, and the rural context. Pap Appl Geogr 2(4):365–382

Haysom G, Battersby J (2023) Urban Food Security and Resilience. In: Resilience and Food Security in a Food Systems Context. Springer International Publishing, Cham, pp 355–388

Hlongwane NH, Vyas-Doorgapersad S (2023) Food security interventions to achieve sustainable development goal two in South Africa. OIDA Int J Sustain Develop 16(02):35–46

Hussain BM, Ryan R, Deierlein AL, Lal S, Bihuniak JD, Parekh N (2023) Food insecurity and health behaviors among a sample of undergraduate students at an urban university. J Hunger Environ Nutr 18(1):65–80

Kawachi I, Kennedy BP (1999) Income inequality and health: pathways and mechanisms. Health Serv Res 34(1 Pt 2):215

Kendall A, Olson CM, Frongillo Jr EA (1996) Relationship of hunger and food insecurity to food availability and consumption. J Am Diet Assoc 96(10):1019–1024

Laborde D, Herforth A, Headey D, de Pee S (2021) COVID-19 pandemic leads to greater depth of unaffordability of healthy and nutrient-adequate diets in low-and middle-income countries. Nat Food 2(7):473–475

Latham MC, Cobos F (1971) The effects of malnutrition on intellectual development and learning. Am J Public Health 61(7):1307–1324

Lemke S, Vorster HH, van Rensburg NJ, Ziche J (2003) Empowered women, social networks and the contribution of qualitative research: broadening our understanding of underlying causes for food and nutrition insecurity. Public Health Nutr 6(8):759–764

Lifland A (2012) Starvation in the Sahel: food security in Africa. Harv Int Rev 34(1):6–7

Mangaung Metropolitan Municipality (2020) Mangaung Metropolitan Municipality spatial development framework (2020–2036) 2022/2023 review. Mangaung Metropolitan Municipality, Bloemfontein

Mbilemi C (2015) Draft climate change adaptation and mitigation strategy for Mangaung Metropolitan Municipality. NM Envirotech Solutions, Bloemfontein

Mkhize X, Mthembu BE, Napier C (2023) Transforming a local food system to address food and nutrition insecurity in an urban informal settlement area: a study in Umlazi Township in Durban, South Africa. J Agri Food Res 12:100565

Mokone SM, Manafe M, Ncube LJ (2023) Healthy eating perceptions of mothers and caregivers of children in South Africa. Health SA Gesondheid 28:2345

Mollier L, Seyler F, Chotte JL, Ringler C (2017) End hunger, achieve food security and improved nutrition and promote sustainable agriculture: SDG 2. A Guide to SDG Interactions: from Science to Implementation. ICSU, Paris, France

Mtembele N (2018) Farmers go hungry after drought. Bloemfontein Courant, 19 November. Farmers go hungry after drought—Bloemfontein Courant (Accessed 6 Apr 2024). https://www.bloemfonteincourant.co.za/farmers-go-hungry-afterdrought/#:~:text=The%20Free%20State%20province%20is,on%20agriculture%20held%20in%20Bloemfontein

News24 (2019) Volunteers help fight against hunger. News24, 24 July. https://www.news24.com/news24/volunteers-help-fight-against-hunger-20190723 (Accessed 9 Apr 2024)

Nkrumah B (2020) We'll fight this little struggle: "Alleviating hunger in South Africa". De Jure Law J 53(1):194–211

Oppelt P (2024) SA's children are dying of hunger: "Malnutrition among Eastern Cape children is at crisis levels. But children are being neglected all over the country by the government." Sunday Times. 30 January 2024. np

Osei RD, Lambon-Quayefio MP (2022) Effects of long-term malnutrition on education outcomes in Ghana: evidence from a panel study. Eur J Dev Res 34(1):1–21

Otekunrin OA (2021) Is Africa ready for the SDG 2 (Zero Hunger) target by 2030? Curr Agri Res J 9(1)

Piaseu N, Belza B, Shell-Duncan B (2004) Less money less food: voices from women in urban poor families in Thailand. Health Care Women Int 25(7):604–619

Roser M (2023) End hunger, achieve food security and improved nutrition and promote sustainable agriculture. Our World in Data. https://ourworldindata.org/sdgs/zero-hunger (Accessed 11 Apr 2024)

Ryan RA, Murphy B, Deierlein AL, Lal S, Parekh N, Bihuniak JD (2022) Food insecurity, associated health behaviors, and academic performance among urban university undergraduate students. J Nutr Educ Behav 54(3):269–275

Sachs JD, Schmidt-Traub G, Mazzucato M, Messner D, Nakicenovic N, Rockström J (2019) Six transformations to achieve the sustainable development goals. Nat Sustain 2(9):805–814

Saint R, Eltaweel A, Adetooto J, Pomponi F, Windapo A (2023) Sandbag housing construction in South Africa: life cycle assessment and operational energy modelling. Int J Life Cycle Assess 28(8):1003–1018

Sasson A (2012) Food security for Africa: an urgent global challenge. Agric Food Secur 1(1):1–16

Stats SA (2021) Focus on food inadequacy and hunger in South Africa in 2021. Stats SA: Pretoria

Shaibu UM (2021) Sustainable development goal 2: assessment of Nigeria's food security situation from 1960–2020. Sustain Dev 21(4):513–518

Smith LC, Haddad LJ (2000) Overcoming child malnutrition in developing countries: past achievements and future choices, vol 30. Int Food Pol Res Inst (IFPRI)

Statistics South Africa (STATSSA) (2023) Sustainable development goals country report – South Africa. Government of South Africa. https://www.statssa.gov.za/MDG/SDG_Country_report.pdf (Accessed 12 Feb 2024)

Tambe BA, Mabapa NS, Mbhatsani HV, Mandiwana TC, Mushaphi LF, Mohlala M, Mbhenyane XG (2023) Household socio-economic determinants of food security in Limpopo Province of South Africa: a cross sectional survey. Agric Food Secur 12(1):19

Todd A (1996) Health inequalities in urban areas: a guide to the literature. Environ Urban 8(2):141–152

Toossi S, Jones JW (2023) The food and nutrition assistance landscape: Fiscal year 2022 annual report

UN DESA (2022) World Population Prospects 2022. United Nations. [Cited 9 May 2023]. https://population.un.org/wpp

Valin H, Hertel T, Leon Bodirsky B, Hasegawa T, Stehfest E (2021) Achieving zero hunger by 2030. A review of quantitative assessments of synergies and tradeoffs amongst the UN sustainable development goals. A paper from the scientific group of the UN food systems summit

Van den Berg L, Raubenheimer J (2015) Food insecurity among students at the university of the Free State, South Africa. South African Journal of Clinical Nutrition 28(4):160–169

Veldhuizen LJ, Giller KE, Oosterveer P, Brouwer ID, Janssen S, van Zanten HH, Slingerland MA (2020) The Missing Middle: connected action on agriculture and nutrition across global, national and local levels to achieve Sustainable Development Goal 2. Glob Food Sec 24:100336

Weiser SD, Leiter K, Bangsberg DR, Butler LM, Percy-de Korte F, Hlanze Z, Phaladze N, Iacopino V, Heisler M (2007) Food insufficiency is associated with high-risk sexual behavior among women in Botswana and Swaziland. PLoS Med 4(10):e260

World Health Organization (2021) The State of Food Security and Nutrition in the World 2021: transforming food systems for food security, improved nutrition and affordable healthy diets for all (Vol. 2021). Food and Agriculture Organization

World Health Organization (WHO) (2022) Barriers to mental health care in Africa. afro.who.int (Accessed 12 Feb 2024)

WHO (2022) Child mortality (under 5 years). https://www.who.int/news-room/fact-sheets/detail/levels-and-trends-inchild-under-5-mortality-in-2020#:~:text=Malnourished%20children%2C%20particularly%20those%20with,under%205%20years%20of%20age

Wright JD, Donley AM, Gualtieri MC, Strickhouser SM (2016) Food deserts: what is the problem? What is the solution? Society 53:171–181

Yang S, Zhao W, Liu Y, Cherubini F, Fu B, Pereira P (2020) Prioritizing sustainable development goals and linking them to ecosystem services: a global expert's knowledge evaluation. Geogr Sustain 1(4):321–330

4 Water and Sanitation for Urban Health: A Gender Perspective on Impacts and Coping Strategies in Mangaung Metropolitan Municipality

Abongile Mgwele, Johannes I. Bhanye, and Kgosi S. Mocwagae

4.1 Introduction

Rapid urbanisation poses significant challenges to urban health, especially in low-income urban areas of the Global South (Abejirinde et al. 2022; Agyei-Mensah et al. 2015). Among the key aspects affecting urban health are access to water and sanitation for all, as outlined in Sustainable Development Goal (SDG) 6. Despite notable progress in increasing access to clean drinking water and sanitation, billions of people in developing countries still lack these fundamental services (Fotio and Nguea 2022). Cholera outbreaks and the COVID-19 pandemic emphasised the importance of sanitation, hygiene, and adequate access to clean water in preventing and containing diseases (Bhanye 2023). Moreover, women and girls, particularly those living in urban poverty, are disproportionately affected by inadequate water and sanitation services (Angoua et al. 2018; Armah et al. 2018). This chapter investigates the gender disparities and implications of inadequate water and sanitation in the Mangaung Metropolitan Municipality. The study further explores the coping strategies employed by women in selected Mangaung communities to address the challenges associated with water and sanitation.

The case study of Bochabelo in Mangaung Metropolitan Municipality is significant. Situated in South Africa, the municipality faces numerous water and sanitation challenges, which makes it an ideal study to understand the impacts on urban health and the coping strategies employed by women. Examining the unique circumstances of Mangaung, the chapter contributes to the existing body of knowledge on urban health, gender disparities, and water and sanitation issues, particularly in the African context. The methodology involves a qualitative survey and key informant interviews to gather qualitative perspectives on women, water, and sanitation, and coping strategies in Mangaung. Additionally, in-depth interviews offered a deeper understanding of women's experiences and coping strategies in relation to water and sanitation challenges. The study also employed observations to have a firsthand experience of women, water, and sanitation in Bochabelo.

A. Mgwele · K. S. Mocwagae
Department of Urban and Regional Planning, University of the Free State, Bloemfontein, South Africa

J. I. Bhanye (✉)
Department of Urban and Regional Planning, University of the Free State, Bloemfontein, South Africa

African Climate and Development Initiative (ACDI), University of Cape Town, Cape Town, South Africa

A. R. Matamanda, V. Nel (eds.), *Sustainable Development Goals and Urban Health*, Sustainable Development Goals Series, https://doi.org/10.1007/978-3-031-68734-1_4

The structure of this chapter is organised as follows. The introduction provides an overview of the research topic, highlighting the significance of water and sanitation for urban health, particularly concerning SDG 6. The subsequent sections explore the conceptual and theoretical approach, contextualise gender and sustainable development in South African, presents results, and discusses the implications of these findings. The chapter concludes by emphasising the need for gender equality, access to water, and sanitation as basic human rights, as well as their role in achieving other SDGs.

4.2 Conceptual and Theoretical Approach

Water, hygiene, and sanitation are fundamental to urban health and sustainable development (UNICEF and WHO 2020). The high demand for water access and supply has, however, resulted in gender inequalities. Therefore, this section explores SDG 6 through the lens of gender equality. It firstly contextualises gender within the constructs of sustainable development. Secondly, it conceptualises SDG 6 from a global context. Thirdly, this section explores the health impacts on women and girls in accessing water and sanitation. Fourthly, the theoretical underpinning of the role of women in water, sanitation, and hygiene (WASH) is discussed.

4.2.1 The Concept of Gender and Sustainable Development

Urban scholars have defined sustainable development from different perspectives. It is a contested concept that intersects with issues of power relations and political economy systems (Nayar 1994; Ogunbanjo et al. 2022). Issues of class, race, governance, and gender in sustainable development are attributed to the divergence of social, economic, and environmental dimensions. In societies, the consideration of gender balances sustainability and improves socio-economic lives for both men and women (Leach 2015). Gender not only targets individual women or men but also holistic gender roles, decision-making, and access to and management of resources (Guzura 2017). Therefore, gender dynamics become anchors in improving and developing societies.

The drivers of development in societies are invested in social norms and systems. Gender, as a socially constructed concept, determines gendered roles and responsibilities (Leach 2015). It becomes a measure of human capabilities to contribute to the development of societies (Ferrant 2015), where women are disadvantaged and marginalised. Therefore, gender inequalities manifest in societies threatening sustainable development. Gender disparities in accessing water and sanitation, for example, are evident from the health implications endured by women.

Furthermore, disparities reveal the marginalised and discriminated women's roles as well as in the need for water, hygiene, and sanitation. Women face more health challenges due to inadequate and unhygienic sanitation facilities and lack of or limited water access (WHO and UNICEF JMP 2023). This health challenge is due to socialisation, where women are primarily responsible for cooking and cleaning. Also, they are more susceptible to infections around menstruation and reproduction (Kayser et al. 2019; Leach 2015; Leach et al. 2016).

Addressing gender inequalities ensures that social justice, inclusivity, and equity are integral to sustainable development (Dugarova 2015). Gender equality means that the roles and needs of women are equally recognised and acknowledged as those of men. Furthermore, gender equity recognises and ensures that women's knowledge, rights, and bodily integrity match the management of resources and decision-making (Leach et al. 2016). Therefore, achieving sustainable development requires the aspects of gender equality to be interlocked with those of the SDGs.

4.2.2 Sustainable Development Goal 6: Ensuring Availability and Sustainable Management of Water and Sanitation for All

The UN General Assembly (2010) recognised access to water and sanitation as a human right and a measure for urban health. As such, SDG 6 ensures the availability and sustainable management of water and sanitation for all by 2030. In reaching this goal, eight targets define the goal with eleven indicators to track the progress worldwide. A focus on urban health and water is largely embedded in targets 6.1–3 being achieved by 2030.

Target 6.1 addresses "…*universal, equitable access to safe and affordable water for all*, thus the proportion of the *population accessing and using safely managed drinking water services*." Complimentary, target 6.3 aims to improve water quality, eliminate pollutants, and improve water treatments. Globally, a stagnant or increased proportion of people still lack access to safe water in urban areas. The SDG 2022 report indicated progress even though 2.2 billion people still did not have access to safe drinking water. The most affected and vulnerable people are those in marginalised and low-income communities. Achieving target 6.1 alone requires an increase of 5 to 8 times the current rate, while little to no progress is made for target 6.3 (UN 2023).

Factors that affect the sustainability of and access to safe water are contingent on the reliability of the water source, transportation and distance to the source, and social affordability and time (Adeyeye et al. 2020; Mocwagae and Mphambukeli 2023). Ideally, the water source should be within the premises and free from contamination (WHO and UNICEF JMP 2023). However, Adeyeye et al. (2020) stipulated that this is far from reality in marginalised and low-income urban areas. Studies in these areas revealed that inadequate water access and management systems exposed people to waterborne diseases (Adelodun et al. 2021; Adeyeye et al. 2020). Poor and old water infrastructure (lack of piped water) has led to contamination from spillages, rust, and pollution. Similarly, drinking water from storage (water tanks, containers, etc.) and water systems (treatment facilities) have been noted as sources of higher levels of contaminants (Zerbo et al. 2021; Adeyeye et al. 2020). Knowledge of domestic water management is also one of the leading causes of water contamination, which leads to diseases and illnesses (Mocwagae and Mphambukeli 2023).

Target 6.2 aims to provide "…*access to adequate and equitable sanitation and hygiene for all and end open defecation, paying special attention to the needs of women, girls, and those in vulnerable situations.*" This is targeted for the population in two ways: access to "…*safely managed sanitation services* and *a hand-washing facility with soap and water.*" In 2022, the United Nations (UN) reported that about 3.4 billion people lacked safely managed sanitation services, and 1.9 billion lacked basic hygiene services. Women and children are the most severely affected, with 1 million deaths reported each year. These deaths are associated with neonatal morbidities and deaths, unclean births, and maternal mortalities (UNICEF 2023).

The WASH concept was introduced as part of target 6.2 to ensure safer water toilets and good hygiene in women and girls. The WASH initiatives are fundamental for access to and provision for adequate water, good hand, and sanitation hygiene. This reduces urban health risks associated with the disposal of human waste, water contamination, and hygiene globally (WHO 2019). However, due to cultural and biological factors and needs, women and girls are disproportionally affected. Societal norms in many developing countries require women to be primarily responsible for water collection and hygiene services (Scherer et al. 2021). This is also exacerbated by women's biological factors, where they require safe and adequate sanitation services for menstruation and reproduction (McCammon et al. 2020). Therefore, adhering to women's needs is essential for targets 6.1–2, to ensure the required "…*six-fold increase for safely managed drinking water, five-fold increase for safely managed sanitation, and a three-fold increase for basic hygiene services*" required by SDG 6.

4.2.3 Women's Health Implications

Lack of access to safe water and hygienic and adequate sanitation poses health threats to women's biological and physical needs. Therefore, women suffer physical and psychological effects from lack of access to water and proper sanitation. This section details these health effects and their implications on achieving SDG 6.

4.2.3.1 Physical Strain and Physiological Effects

Many low-income households need to walk or travel long distances to access water. Water sources outside the household premises or not readily available require women and girls to travel distances for access (Tomberge et al. 2021). The long distances[1] have resulted in women and girls experiencing physically demanding tasks. Therefore, compared to men, women and girls struggle with head, neck, and thoracic pain (Geere et al. 2010). This increases fatigue and affects daily functioning. Some women have been physically injured during the long distances travelled to access water like bone fractures, bruises, and dislocations (Sharma and Singh 2012; Venkataramanan et al. 2020).

Inadequate WASH systems and facilities also pose detrimental physiological effects on women and girls. Higher illnesses and deaths due to contaminated water and unhygienic shared toilets have been reported. Studies by Prüss-Ustün et al. (2019) and Ademas et al. (2020) reported higher disability-adjusted life years (DALYs) due to illnesses and diseases. Women are at risk of several parasitic infections, causing diarrhoea, maternal anaemia, and reproductive tract infections (Ademas et al. 2020; UNICEF 2023). Compared to men, accessing safe water and adequate WASH systems has resulted in physical and physiological health effects and affected their psychological states.

4.2.3.2 Psychological Effects on Women and Girls

Reports on gender-based violence or intimate partner violence have suggested that lack of access to WASH as a contributing factor. The long distances from water sources have limited household water availability and delayed return for other household duties like cooking and cleaning. According to Mushavi et al. (2020), this may reduce partners' trust and lead to violence. Furthermore, the failure to complete other socially contracted roles for women results in physical abuse (Choudhary et al. 2020). Physical abuse from outside the household has also been reported. Kayser et al. (2019) and the WHO (2021) stated that women and girls are at a higher risk of being harassed and raped by men when fetching water and sharing toilets. Therefore, access to WASH is more than just a right for women, but also about personal safety. Shared facilities cause stress and anxiety among young girls and women, particularly during their menstrual periods (Saleem et al. 2019).

Apart from sexual harassment and abuse, one in five girls of primary school age miss out on schooling during their menstrual periods, because of limited water and hygiene at school and home. The social expectations for girls to fetch water, cook, and clean result in them not having enough time for school. The lack of education has contributed to a high dropout for girls (Shah et al. 2022). This negatively affects women's empowerment and equity and poses a challenge for sustainable development.

4.2.4 The Role of Women in Achieving SDG 6, a Theoretical Approach

Women are often excluded from urban-related decision-making bodies and resources. Consequently, their role in accessing WASH and participation in water service design and governance is marginalised. Therefore, this section investigates the role that women can play through a feminist urban theory. Feminist thought, in urban studies, originated in the 1970s through analysing the urban form concerning women. It was founded on the constraints endured by women and exposed the differentiated gender roles that continued to economically, socially, and spatially marginalise women (Sandercock

[1] A long distance is more than 30 minutes from the water source, including round trips, lining up and waiting, and more than 1 km. (UN 2022)

and Forsyth 1992). Due to the limitations of focusing on women alone, the modern feminist urban theory advanced to use a gendered view and its inter-relations. The feminist approach reveals in-depth insights into urban spaces and everyday spatial practices that are inherently gendered (Beebeejaun 2017). It challenges the gendered forms of spatial oppression, marginalisation, and stereotyping.

Women continue to experience urban spaces that are still predominantly patriarchal (Rampaul and Magidimisha-Chipungu 2022). Spatial urban practices in accessing WASH have revealed that women's roles and needs are oppressed and marginalised. Therefore, achieving SDG 6 requires the current roles of women to be challenged, hence the feminist urban approach. The feminist perspective proposes women as agents of power and change (Allen 2005). It further improves women's lives in their existing roles and empowers them to adopt new roles (Leach et al. 2016). In theorising women's roles in accessing WASH, we discuss three approaches. Firstly, challenging social norms and gender stereotypes around women's spatial roles and experiences. Secondly, change in structures of power in water governance. Thirdly, increase the availability of gender-disaggregated WASH data.

Social and cultural norms that are embedded in bodily conduct undermine the WASH initiatives for women. This is evident in the works of Young (2005, 2008), where she argues that women's embodiment has resulted in them continuously hiding and confined during their menstrual cycle. This can be traced back to historical and biblical norms, where women were considered unclean (Witherington III and Witherington 1987). In marginalised communities that do not have access to WASH facilities inside their premises, social norms become superior to women's needs. The female body alone determines the spatial needs to maintain hygiene and access safe toilets and water, which differ from men's needs in the same area. Similarly, gender stereotyping on the role of women in collecting water, cooking, and cleaning limits women from adopting any other opportunities. According to Jalali (2021), this embodiment and stereotyping reveal the gendered nature of WASH deprivation and power structure. Therefore, challenging the social norms of discriminating against women and girls' bodies, roles, and needs becomes critical in accessing WASH.

Health implications reflect the gendered complexities and inequalities of WASH and the power dynamics embedded in them (Carrard et al. 2013; UNICEF 2015). Even with rich literature on health implications for women, their voices remain unheard. Policies and strategies do not reflect realities on the ground, uncovering the unequal power relations. Structures of power dominate formal institutional policies and legislations that govern collective institutional practices of assessing WASH (Kabeer 2018). The institutional arrangements remain ineffective and limited in the provision for WASH (Wardeh and Marques 2023). These are embedded in the patriarchal ideology and culture that oppresses and discriminates against women. The feminist urban theory thus advocates for exposing and dismantling these power structures to create gender-inclusive and equal opportunities and urban experiences (Allen 2005). Women will then be able to able to form part of and drive water governance initiatives by dismantling these power structures in urban areas.

The role of women in accessing WASH can also be determined through the increased gender-sensitive data. Increasing gender-disaggregated WASH data entails acknowledging that gender relations are specific to communities. Historical and political differences also account for spatial practices. Therefore, data on WASH needs to encompass this to allow the restructuring of gender roles (Caruso et al. 2021). This is seen in marginalised communities where women experience adverse inequalities compared to those in affluent areas (Rossouw and Ross 2021). Gendered data gives women voices and combines different experiences and perspectives of WASH in different places. This gives agency to women's water needs and shifts policies and practices towards more localised or contextual WASH interventions. Furthermore, it allows recovery strategies specific to disadvantaged girls and women in accessing water.

The role of women in achieving SDG 6 should be seen through the feminist lens. Feminist urban theory and the SDG 6 are interlocked in that they both seek to ensure gender equality. Women and girls should have equitable access to hygiene, water, and sanitation. Through challenging social norms, dismantling power dynamics, and increasing gender segregation, the role of women and girls, particularly in vulnerable situations, is recognised.

4.3 Sustainable Development Goal 6 and Urban Health in South Africa

The South African Constitution states that everyone has basic human rights, which includes protection from a harmful environment with access to safe water and sanitation. Section 24(a) of the Bill of Rights states "*to an environment that is not harmful to their health or well-being*". Section 27 of the Bill of Rights further states "*Everyone has the right to have access to— (a) health care services, including reproductive health care; (b) sufficient food and water*". Aligned with the Constitution, the NDP 2030 advocates for effective and sustainable management of water access and sanitation (NPC 2012). The NDP emphasises that access to safe water and sanitation is significant for urban health, economic progress, and social cohesion. The Bill of Rights and NDP further drive the adoption of water and sanitation policies and legislations, notably the National Water and Sanitation Master Plan, the Water Service Act of 1997, and the National Water Act of 1998. These legal frameworks guide the implementation of effective water supply and sanitation by 2030.

However, the challenge of water security or scarcity delays the success of ensuring effective water supply and sanitation (Muller et al. 2009). Inadequate infrastructure, water governance, and financial investment further exacerbate the challenge (RSA 2018). Inequalities through gender, class, and race also play a significant role in delaying the achievement of SDG 6 in South Africa (NPC 2012). This is evident in water policies and governance systems and the lack of skilled water professionals in the country. As a result, about three million South Africans still lack access to basic water, and 14 million lack access to safe and adequate sanitation (RSA, DWS 2021).

Over time, progress has been made to ensure everyone has access to water and sanitation facilities. This is evident from the "*free basic water*" and "*free to subsidy for sewage and sanitation*" initiatives directed at poor households..[2] South Africa is a member state of the UN and abides by the provision prescripts that water is considered accessible if within a distance of 1 km, with a volume of 50–100 l, and not more than 30 minutes away, and should not exceed total household expenses by 3% (Mocwagae and Mphambukeli 2023). Based on the UN water provision prescripts, poor households receive 6000 l of water per month per household. According to the Stats SA General Household Survey (GHS) (2018), there has been an increase from 50.8% in 2009 to 59.1% in 2018 of people who did not pay for water. In contrast, Scheihing et al. (2020) argue that the free water initiative is not financially feasible and still requires an improved management system. Therefore, a strategy to enhance water management, particularly in poor households, is key.

Access to water is reported to be steadily improving nationally in the past five years. In 2021, 88.7% of households have access to an improved water[3] source off and on-site. A larger percentage, 45.2%, is attributed to water access through piped (tap) water in dwellings (Stats SA, GHS 2021). While the access to WASH facilities is approximately 73.5%. Even though nationally, there have been some improvements toward achieving SDG 6, the percentage in the Free State province has decreased from 96.9% in 2012 to 93.6% in 2021. Figure 4.1 depicts the percentage of household access to piped or tap water nationally and in the Free State. This

[2] According to Stats SA, as of 2022, a person receiving an amount of R1417 or less per month is considered poor.

[3] Improved water is accessed through water pipes or taps, Stats SA, 2021.

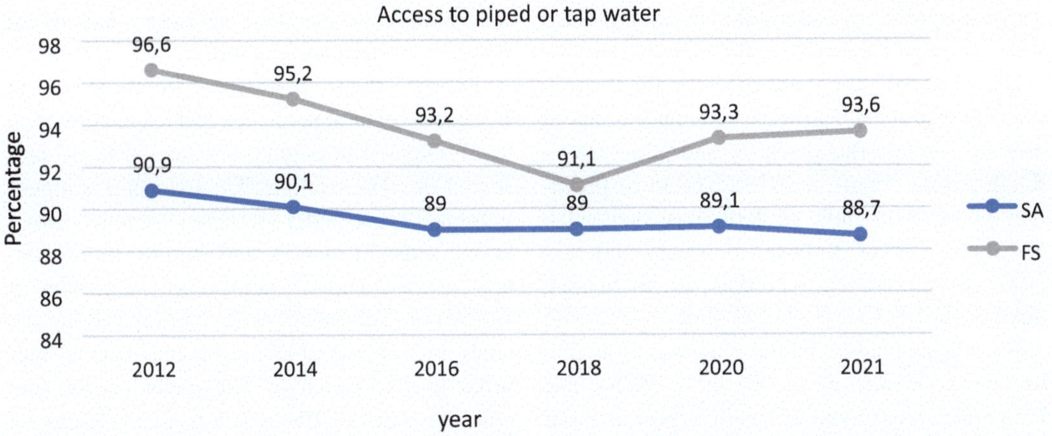

Fig. 4.1 Percentage of households with access to water (SA: South Africa; FS: Free State province). (Source: data compiled from Stats SA GHS report 2021 (https://www.statssa.gov.za/publications/P0318/P03182021.pdf))

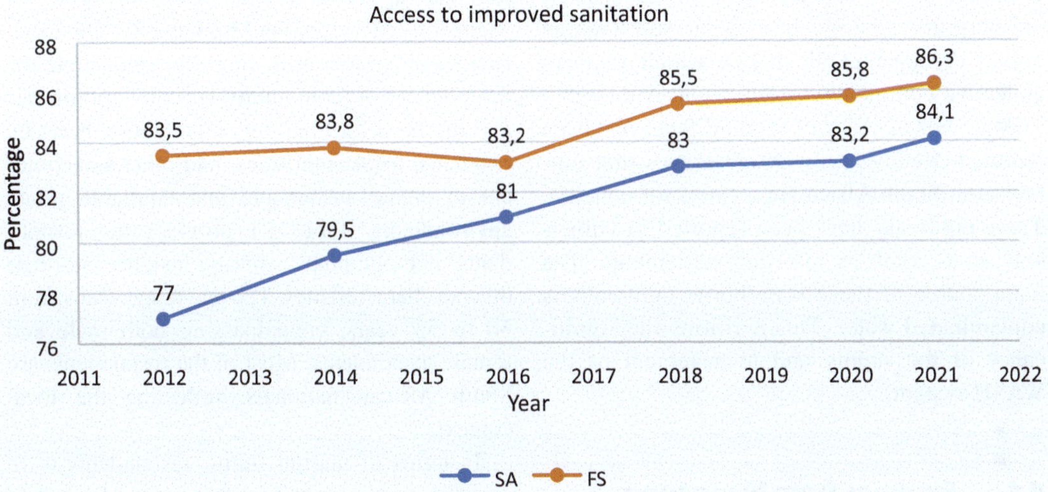

Fig. 4.2 Household percentage with improved sanitation (SA: South Africa; FS: Free State province). (Source: data compiled from Stats SA GHS report 2021 (https://www.statssa.gov.za/publications/P0318/P03182021.pdf))

shows that the province is performing above the national mark, although with a declining percentage. The notable decrease has led to some households using alternate sources for accessing water. In some parts of the country, a concept of water shedding, where there are regular water interruptions, has been introduced (Wurster 2020).

The limited access to water in South Africa has also challenged hygiene and sanitation initiatives. South Africa recorded about 84% of households that have access to either flushed toilets or ventilated pits (improved sanitation) in 2021. This shows a slow improvement compared to the past five years, as depicted in Fig. 4.2. However, about 16% of households depend on pit latrines without ventilation, chemical toilets, or other means (Stats SA 2021). In the Department of Water and Sanitation Report to Parliament, 2021/2022, it is reported that 289,791 households practice open defecation. Inadequate sanitation facilities and limited water access cause health complications in South African women, particularly those in low-income and rural communities.

The inadequacy has also resulted in contaminated water, posing health risks in these communities.

The limited and inadequate water, sanitation, and hygiene facilities account for about 88% of waterborne diseases and illnesses in South Africa (Hemson 2016). The most prevalent diseases include cholera, gastroenteritis, and typhoid fever. Between February and July 2023, South Africa recorded an increasing number of 1073 suspected cholera cases, with 198 confirmed cases. Females account for 52% of these confirmed cases (SA news 2023). The Free State Department of Health reported about four fatalities due to cholera. Similarly, contaminated water sources cause gastroenteritis particularly in children, and remains a health challenge. It has been noted as the highest cause of mortality among children in sub-Saharan Africa (Troeger et al. 2018), which requires collective continental intervention. Although a low number of cases of typhoid fever has been reported, Keddy et al. (2018) argues that outbreaks in South Africa are a cause for concern. These outbreaks have been reported in vulnerable areas, such as informal settlements. The main source of these waterborne outbreaks is contaminated water. This reaffirms the significance of the supply and management of the WASH system.

4.4 Findings from Mangaung Metropolitan Municipality: Bochabela

4.4.1 Study Demographics and a Brief Background of Bochabela

Bochabela is a township located in Mangaung Metropolitan Municipality, in the Free State province of South Africa. Bochabela, like many townships in South Africa, has a rich history shaped by the country's complex past, including the Apartheid era. A mix of residential areas, informal settlements, and community facilities characterises the township. Bochabela has faced challenges such as inadequate infrastructure, limited access to essential services, and issues related to water and sanitation.

Despite these challenges, communities in Bochabela exhibit resilience and resourcefulness, as highlighted in coping strategies that are presented in other sections. The township is a geographical space and a vibrant community with a strong sense of identity and culture. In the context of water and sanitation, the experiences of women in Bochabela who participated in the study, provide valuable insights into their broader urban health challenges. Understanding the specific dynamics of Bochabela residents is crucial for designing effective interventions and policies, which address their needs. As efforts continue to improve water and sanitation services, Bochabela remains a dynamic and evolving part of Mangaung Metropolitan Municipality. The experiences of women in Bochabela demonstrate the importance of community-specific approaches and the need for ongoing collaboration between residents, local authorities, and other stakeholders to create sustainable and equitable urban environments. Table 4.1 provides the respondents' demographics, offering insights into their diverse characteristics. The age range spans from 30 to 53 years, encompassing both male and female respondents. Most of the respondents are South African nationals, reflecting the local context.

In terms of marital status, respondents were married, single, or had a life partner. Household sizes varied from 2 to 7 members. Respondents' duration of residing in Bochabelo range from five years to their entire lives. Employment status varied across the respondents, including self-employed individuals, street vendors, subsistence farmers, and those seeking employment. Educational backgrounds also differed, with respondents having a primary school education to holding diplomas and certificates. Homeownership situations were diverse, with some respondents owning their homes through inheritance or personal efforts while others were renting. The respondents' average monthly income also varied, providing an overview of the economic diversity within the community. This demographic overview sets the foundation for

Table 4.1 Demographics of respondents

Age	Gender	Nationality	Marital status	Household size	Years staying in the suburb	Employment status	Level of education	Home ownership situation	Average monthly income (ZAR)
46	Male	South Africa	Have life partner	3	7	Self-employed	Certificate	Inheritance	4000–5000
40	Male	South Africa	Married	3	5	SAB	Carpentry certificate	Renting	8000
53	Female	South Africa	Married	3	25	Unemployed	Matric	Ownership	<4000
35	Female	South Africa	Married	4	8	Subsistence farmer	Matric	Inheritance	–
49	Female	South Africa	Single	2	All my life	Unemployed / street vendor	Standard 6	Renting	–
33	Female	South Africa	Single	2	All my life	Hairdresser	Grade 11	Ownership/ inheritance	2500
35	Male	South Africa	Single	6	All my life	Car sales agent	Matric	Lives home	
32	Male	South Africa	Married	4	All my life	Seeking employment	Standard 7	Inheritance	2000
39	Male	South Africa	Single	7	All my life	Unemployed (selling alcohol)	Diploma	Ownership	2000–4000
30	Female	South Africa	Married	4	All my life	Seeking employment	Matric	Renting	2000

understanding the perspectives and experiences of women regarding water and sanitation challenges in their urban context.

4.4.2 Water and Sanitation Realities in Bochabelo

The findings into the water and sanitation situation in Bochabelo revealed both positive aspects and significant challenges. Regarding water services, respondents generally reported accessibility and availability within their households. Notably, water is not prepaid but billed after use. However, water quality poses a concern, with instances of it being reported as dirty and occasionally undrinkable. Respondents expressed sporadic water cuts and shortages, particularly concerning for households with toddlers, leading to the purchase of store-bought water for drinking. One respondent articulated this concern, "Water is good, but sometimes it is undrinkable. We get a bill after using it, and sometimes it is brown. So, we buy store water for the kids."

Various sanitation facilities were reported, including indoor water systems, toilets, outdoor flush toilets, and shared toilets for multiple families. Despite the provision of these facilities, challenges persist with the sewage system. Instances of clogging, leaks, and overflows were noted, with delayed responses from the municipality in addressing these issues. This resulted in unpleasant smells and compromised hygiene in the affected areas. A respondent shared their experience, "We have an indoor toilet, but sometimes the sewage system leaks. It takes a

Source: Fieldwork

Fig. 4.3 A makeshift toilet in an informal settlement in Bochabela. (Source: Fieldwork)

while for the municipality to fix it, and the smell can be horrible." Figure 4.3 shows a makeshift toilet in an informal settlement in Bochabela, threatening the health of the people in the community.

The issue of garbage collection and disposal adds another layer of complexity. Municipal garbage collection was reported to be inconsistent, prompting residents to either dispose of garbage in vacant sites or burning it at dumping sites. The irregularity of collection contributed to challenges in maintaining a clean environment. A respondent conveyed this challenge, "Garbage collection is not consistent. We sometimes dump it at a vacant site.

When it is not collected, it is burned at the dumping site, creating a bad smell." Figure 4.4 shows garbage dumped all everywhere in Bochabela community.

The presence of shared community facilities emerged as a recurring theme. Shared taps and toilets were mentioned, with some instances of multiple families sharing the same flush toilet. While water is generally available in the community, infrastructure maintenance challenges were highlighted, particularly broken toilets requiring repairs. One respondent detailed their situation, "We share toilets in the yard, and one needs fixing. All families use the same toilet. Sewage sometimes overflows, but the municipality fixes it."

4.4.3 Variations in Urban Realities: Contrasting Neighbourhood Dynamics in Mangaung Metropolitan Municipality

The study also revealed the disparities among different neighbourhoods within the Mangaung Metropolitan Municipality, shedding light on residents' diverse water and sanitation experiences. In more affluent neighbourhoods, respon-

Fig. 4.4 Garbage dumped everywhere in a Bochabela community. (Source: Fieldwork)

dents highlighted the municipality's consistent garbage collection, and minimal instances of sewage leakages or clogging. A respondent emphasised, "In the suburbs, there is rarely ever an issue. The municipality collects garbage regularly, and we hardly experience sewage problems here." A stark contrast emerged in informal settlements locally referred to as *dikoereng*. Here, the municipality does not extend its garbage collection services. One respondent revealed, "I know for a fact that they do not collect any garbage in the informal settlements. It is a different reality there."

The water challenge, however, was deemed a South African problem, affecting various areas. Despite this, respondents noted that more affluent neighbourhoods exhibited a well-functioning garbage collection system and lack of sewage leakages. A respondent stated, "Water issues are everywhere, but you never see garbage bags in the suburbs. They have a functioning collection system, and there are no leakages." Certain neighbourhoods, particularly those without government subsidised (RDP) houses, face significant challenges, lacking both water and sewer systems. A respondent highlighted, "Some neighbourhoods do not have water or a sewer system, especially where RDP houses have not been built yet." In contrast, suburbs enjoy more consistent services, with garbage collected regularly. When asked about the willingness to pay for garbage collection, a resident responded affirmatively, with a budget cap of R10. "I would pay for it, but nothing above R10. In suburbs, it is done well."

Disparities persists, especially in informal settlements, where running water is absent and garbage goes uncollected (Fig. 4.5). A resident articulated the struggle, "Here, it is collected, but the consistency needs improvement. Informal settlements face more significant challenges." Some neighbourhoods praised the municipality's consistency, emphasising that residents do not require elevated frames for garbage disposal. They trust the scheduled collection days. "In

Fig. 4.5 An informal settlement in Bochabela where running water is absent and garbage goes uncollected. (Source: Fieldwork)

most neighbourhoods, garbage is collected consistently on a particular day. That does not happen here. The municipality takes forever," a resident shared.

The Bochabela region showcased significant differences within itself. On one side, residents experience the frustration of delayed garbage collection, leading to road blockades. On the other side, a permanent dumping site exists near Pelonomi Hospital, revealing the dichotomy within Bochabela. Despite varied neighbourhood dynamics, respondents noted a commonality in water quality across the municipality, with occasional drinkability issues. The need for more consistent garbage collection emerged as a shared concern, prompting discussions on potential solutions. A respondent proposed, "In some neighbourhoods, residents have come together, paid a certain amount, and hired a private company for waste collection. That could work here in Bochabela, but it must be relatively inexpensive."

4.4.4 Impact of Water and Sanitation Quality on Women's Health in Bochabela

Exploring the nexus between water and sanitation services and women's health in Bochabela revealed interesting insights into the gender-neutral but communal implications of these essential services. Contrary to expectations of gender disparities, a respondent noted that both men and women faced health challenges during a prolonged period of dirty water in Mangaung. "There was a time when the water in Mangaung was dirty for a long time, and both women and men who drank it fell ill. It was not one gender suffering more than the other," the respondent emphasised, highlighting the community's shared vulnerability. In contrast, another respondent asserted that women in their community did not fall ill due to water and sanitation-related issues. "No. Women get sick, but from illnesses that are unrelated to water and sanitation," the respondent clarified, emphasising a distinction between gender-neutral health issues and those directly linked to water and sanitation. The perceived health risks were minimal for some communities, particularly those with flush toilets and running water. A respondent noted, "Here, everyone has a flush toilet and running water in their yard. Such problems exist in *dikoereng*/informal settlements with no water or toilets."

However, challenges arose in shared facilities, impacting women's health differently. In a yard where residents shared a toilet, women faced privacy and hygiene issues. "The only complaint women have in the yard is the fact that they have to share the toilet outside with men. This leaves little room for privacy, and the men leave it dirty. The toilet is also used by strangers here to visit," a respondent explained, shedding light on the specific challenges women face in communal sanitation spaces. The health risks were more pronounced in informal settlements, where basic water and sanitation services were lacking. Another respondent stated, "The only health risks that occur are at informal settlements where they do not have running water or toilets. Here, such problems do not exist."

Women reported minimal health issues in specific neighbourhoods with consistent access to water and sanitation services. "In this neighbourhood, women are unaffected because they can access the same services as men. They have running water in their houses, and garbage is sometimes collected or dumped at a vacant site. So, it is not like they live in their filth," a respondent shared, emphasising the positive impact of adequate services on women's health. However, a poignant case emerged where a shared toilet compromised women's health due to the lack of privacy. The respondent narrated a personal experience, stating, "The shared toilet means no privacy for women. I had a rash in my groin area, and when I went to the doctor, they said the bacteria from the toilet seat most likely caused it." This anecdote revealed the intersectionality of water and sanitation services with women's health, emphasising the need for solutions that consider privacy and hygiene in shared facilities.

4.4.5 Gender Dynamics and Power Relations in Water and Sanitation: A Bochabela Community Perspective

Exploring the interplay between gender dynamics, power relations, and water and sanitation services within the Bochabela community offered an understanding of the equality and challenges experienced by residents. Contrary to expectations of gender-based disparities, a prevailing sentiment among respondents was that gender equality was upheld in their communities. "Gender disparities do not exist. All genders are equal in this community. All households receive the same services and complain about the same disservices," a respondent emphasised, challenging the notion of gender-specific disparities. Several respondents echoed this sentiment, asserting that gender dynamics did not play a role in determining access to water and sanitation. "Such dynamics do not exist in the community. All genders get water and sanitation," one respondent stated, emphasising the universality of service provision regardless of gender.

However, a distinct perspective emerged when examining power dynamics within the community. While gender did not seem to influence access to water and sanitation directly, power imbalances were noted in the community's waste management. A respondent explained, "The only power dynamics that affect us as a community are that the garbage is not collected in time. This causes a problem of waste disposal. The sewage leaks are also not attended to quickly enough, making it unsafe for the kids because they play in the streets." This highlights a different dimension of power dynamics linked to the timely management of waste and sewage issues affecting the community's overall well-being.

In contrast, respondents asserted that power dynamics were absent when it came to water and sanitation services controlled by the municipality. "There are no power dynamics when it comes to water and sanitation because it is provided and controlled by the municipality," one respondent clarified, underscoring the centralised authority in service provision. The theme of centralised authority continued to dominate the discussions, with respondents consistently asserting that no power dynamics were associated with water and sanitation in their communities. "There are no power dynamics because there is one provider of water and sanitation," a respondent reiterated, emphasising the singular source of service provision. Furthermore, the role of the government in service provision was highlighted, with respondents expressing that the absence of power dynamics was attributed to governmental responsibility. "There are no community power dynamics because the government provides the services," one respondent noted, reinforcing the idea that a centralised authority contributed to equal service distribution.

Interestingly, while the respondents largely dismissed the influence of power dynamics in water and sanitation access, a unique case emerged. A respondent noted, "It is not the power dynamics that affect sanitation for women, but the landlords decide to have one toilet that accommodates both men and women in the yard. The overcrowding makes it worse." This exception sheds light on the localised impact of decisions made by landlords, influencing the gendered experience of sanitation within certain households.

4.4.6 Coping Strategies in the Face of Water and Sanitation Inadequacies: Insights from Women in Bochabela

The study also explored the resourceful coping strategies employed by women in Bochabelo to navigate the challenges associated with water and sanitation inadequacies, shedding light on their resilience and adaptability. In instances where water scarcity is a concern, respondents revealed the common practice of storing water in 20-litre buckets. The stored water serves as a contingency for cooking, bathing, and household chores during periods when water is not readily available. "Even though water seldom runs out, water is stored in 20-litre water buckets for when it does happen. This will be the water used for cooking, bathing, and household chores," a respondent disclosed, emphasising the proactive approach taken by residents. Garbage disposal strategies emerged as a crucial aspect of community coping mechanisms. Residents resort to throwing away garbage at an informal dumping site near Pelonomi Hospital when faced with challenges in municipal collection services. Another respondent explained, "The strategy for garbage is to throw it at an informal dumping site close to Pelonomi Hospital." This decentralised waste management approach reflects the community's resilience in the face of inconsistent municipal services.

Preparation for potential water shortages involves storing water in containers, including jerry cans. Boiling water before consumption and storage in jerry cans for emergency use during shortages has become a common practice. "Preparation for a water shortage is to have buckets of water. Water used for cooking and drinking is boiled and put in jerry cans," a respondent elaborated, illustrating the proactive measures taken to ensure a continuous water supply. The storage of water in drums emerged as a prevalent coping strategy. Respondents highlighted the importance of having a drum filled with water in case of emergency shortages. In some cases, respondents mentioned using public toilets in town when necessary, showcasing adaptive strategies beyond the household level. "There is always a drum filled with water in case there is a shortage. I also use the toilets in town close to where I'm selling," a respondent shared, revealing a dynamic approach to addressing water challenges.

Solid waste management strategies included burning garbage at the dump site when municipal collection services were unreliable. The use of elevated metal structures to safeguard garbage from dogs at night demonstrated a thoughtful approach to waste disposal. "The garbage is put in a metal structure that is alleviated off the ground so that dogs do not open the garbage bag during the night," a respondent explained, reflecting the community's resourcefulness in waste management.

Shared toilets posed specific challenges for some women, leading to innovative coping strategies. Respondents mentioned avoiding the shared toilets whenever possible and opting for alternative facilities. The respondent said that "when she finds a piece job, she goes to the toilet there, and the kids go to the toilet at school. They avoid using the shared toilet as much as possible," highlighting the personal measures taken to mitigate discomfort and maintain a sense of privacy.

4.5 Discussion and Policy Implications: Towards Gender-Inclusive Water, Sanitation, and Hygiene (Wash) Initiatives

In summary, the study provides a comprehensive and nuanced understanding of water and sanitation services in the Mangaung Metropolitan Municipality, Bochabela community. The findings highlight the importance of addressing both positive aspects and challenges related to water and sanitation to enhance urban health outcomes, focusing on women in these communities. The need for improved maintenance, consistent services, and community-specific interventions emerges as crucial for fostering a healthier and more sustainable living environment. The diverse challenges and successes experienced by residents in Bochabela reveal the necessity for a community-specific approach to tackle urban disparities in water and sanitation services. While the findings suggest a predominantly egalitarian landscape, the influence of power dynamics in waste management and localised decisions, such as shared toilets, emphasises the importance of considering both community and household-level factors in shaping the water and sanitation experiences of residents, particularly women.

The coping strategies employed by women in Bochabela reflect their adaptability and resilience in the face of water and sanitation challenges. The findings illuminate the community's proactive approach to ensuring basic needs are met despite systemic inadequacies, from innovative water storage practices to creative waste disposal methods and solutions for shared toilets. What is essential is tailored interventions that address privacy concerns and promote gender-inclusive health outcomes in urban communities, emphasising a holistic and community-specific approach to urban health and sustainable development.

From the study, gender perspectives are integral in balancing the socio-economic dimensions of sustainable development. Women, often marginalised and disadvantaged, bear the brunt of gender inequalities manifested in limited access to water and sanitation. The socio-cultural norms shaping gender roles intertwine with issues of WASH, highlighting a holistic approach to sustainable development, which addresses gender dynamics at their core. While SDG 6 aims to ensure universal access to safe water and sanitation by 2030, the study's findings illuminate the persistent challenges in achieving this goal. The disparities in water access and sanitation facilities, particularly in marginalised communities, exemplify the urgency of targeted interventions. The water crisis and inadequate sanitation pose significant health risks, again emphasising holistic and gender-specific strategies.

The health implications on women in this study, reveal the physical and psychological toll of inadequate WASH services. The physical strain of accessing water and the physiological effects of contaminated facilities call for improved infrastructure. Additionally, the study highlights the psychological impact, including gender-based violence and the societal expectations affecting education and empowerment. Feminist urban theory provided a lens to examine and challenge existing power structures contributing to gender disparities in WASH. The chapter explores three key approaches: challenging social norms and gender stereotypes, dismantling power dynamics in water governance, and advocating for gender-disaggregated WASH data. Empowering women as change agents is fundamental to achieving SDG 6 through the feminist urban perspective. South Africa's constitutional commitment to water and sanitation as fundamental human rights aligns with the NDP 2030. However, challenges persist, including water scarcity, inadequate infrastructure, and gender, class, and race-based inequalities. While progress has been made, the study's findings emphasise targeted interventions to ensure sustained access to water and sanitation, particularly in marginalised communities.

The outlined policy implications provide a roadmap for policymakers to create gender-inclusive, sustainable, and effective WASH initiatives, ultimately contributing to the realisation of SDG 6 in South Africa and beyond.

- **Gender-responsive WASH policies:** The study emphasises the need for policies explicitly designed to address gender disparities in water and sanitation. Gender-responsive WASH initiatives should challenge societal norms, ensure equitable access, and prioritise the specific needs of women and girls.
- **Infrastructure development:** The findings highlight the critical role of infrastructure in shaping WASH outcomes. Policymakers should prioritise investments in water and sanitation infrastructure, especially in marginalised communities, to alleviate the physical burden on women and improve overall health outcomes.
- **Feminist urban approaches in governance:** Adopting feminist urban approaches in water governance is imperative. This involves dismantling patriarchal power structures, promoting women's active participation in decision-making processes, and challenging spatial practices that marginalise women in accessing WASH services.
- **Gender-disaggregated data collection:** Policymakers should prioritise collecting and analysing gender-disaggregated data related to water and sanitation. This approach ensures a nuanced understanding of gender-specific challenges, enabling the development of targeted interventions that address the unique needs of women and girls.
- **Community engagement and empowerment:** Policies should foster community engagement, empowering women to actively participate in the design, implementation, and monitoring of WASH initiatives. This community-centric approach ensures sustainable solutions that resonate with local contexts.

4.6 Conclusion

In conclusion, this in-depth exploration of water and sanitation services in the Mangaung Metropolitan Municipality based on the case study of Bochabela revealed the complex dynamics that significantly impact urban health, particularly for women. The study, situated within the context of rapid urbanisation and SDG 6, highlighted the persistent challenges low-income urban areas face, especially in the Global South. Despite substantial progress, many communities, predominantly in developing countries, still lack access to clean drinking water and sanitation.

The recent Cholera outbreaks and the COVID-19 pandemic highlighted the pivotal role of safe environments, emphasising the importance of sanitation, hygiene, and access to clean water in preventing and containing diseases. Women and girls, particularly those struggling with urban poverty, bear a disproportionate burden when it comes to accessing water and sanitation. This chapter revealed the gender disparities and the implications of inadequate water and sanitation in Mangaung Metropolitan Municipality, shedding light on women's lived experiences in the Bochabela community. The comprehensive literature review of the study established a robust theoretical and conceptual foundation, framing the study within the broader context of SDGs and gender dynamics. It illuminates how gender inequalities manifest in water and sanitation access, subsequently affecting women's health. The physical and psychological health implications highlight the urgency of prioritising gender-sensitive strategies in achieving SDG 6 targets. The discussion further explored coping strategies employed by women, showcasing their adaptability and resilience in the face of systemic inadequacies and providing practical insights for sustainable interventions.

The study emphasises the critical need for a holistic approach to address urban health disparities, particularly water and sanitation. The interplay of gender dynamics, power relations, and community-specific factors highlight the necessity for tailored interventions. While certain neighbourhoods exhibit predominantly egalitarian water and sanitation services, the influence of power dynamics, especially in waste management and shared toilet decisions, unveils the importance of recognising both community and household-level factors in shaping the experiences of residents, with a particular focus on women. The coping mechanisms, ranging from water storage practices to innovative waste disposal methods and solutions for shared toilets, reveal the community's proactive engagement in meeting basic needs. These coping strategies serve as a testament to the resourcefulness of women in navigating challenges related to water and sanitation, underlining the potential for community-driven solutions.

The broader contribution of this study lies in advocating for gender equality and framing access to water and sanitation as fundamental human rights crucial for achieving various SDGs. The study recommends prioritising women's participation in water governance, emphasising the need for sex-disaggregated data on water, sanitation, and hygiene, challenging social norms around women's leadership, and addressing gender-based violence in water and sanitation issues. These recommendations serve as a foundation for future research and interventions to create a more inclusive and equitable urban environment for all. Future research needs exploration into the evolving dynamics of water and sanitation services, with a specific focus on the intersectionality of gender, class, and race. Additionally, research endeavours should explore the effectiveness of community-specific interventions and their long-term impact on urban health outcomes. As urban landscapes continue to transform, ongoing studies will be vital in informing responsive policies and practices, ensuring sustainable development and improved well-being for urban populations, with a particular emphasis on the most vulnerable.

References

Abejirinde IOO, Gwacham-Anisiobi U, Affun-Adegbulu C, Vanhamel J, Van Belle S, Marchal B (2022) A perspective on urban health systems and research for equitable healthcare in Africa. BMJ Glob Health 7(9):e010333

Adelodun B, Ajibade FO, Ighalo JO, Odey G, Ibrahim RG, Kareem KY, Bakare HO, Tiamiyu AO, Ajibade TF, Abdulkadir TS, Adeniran KA (2021) Assessment of socioeconomic inequality based on virus-contaminated water usage in developing countries: a review. Environ Res 192:110309

Ademas A, Adane M, Sisay T, Kloos H, Eneyew B, Keleb A, Alemu K (2020) Does menstrual hygiene management and water, sanitation, and hygiene predict reproductive tract infections among reproductive women in urban areas in Ethiopia? PLoS One 15(8):e0237696

Adeyeye K, Gibberd J, Chakwizira J (2020) Water marginality in rural and peri-urban communities. J Clean Prod 273:122594

Agyei-Mensah S, Owusu G, Wrigley-Asante C (2015) Urban health in Africa: looking beyond the MDGs. Int Dev Plan Rev 37(1):53–60

Allen A (2005) Feminist perspectives on power. https://plato.stanford.edu/ENTRIES/feminist-power/

Angoua ELE, Dongo K, Templeton MR, Zinsstag J, Bonfoh B (2018) Barriers to access improved water and sanitation in poor peri-urban settlements of Abidjan, Côte d'Ivoire. PLoS One 13(8):e0202928

Armah FA, Ekumah B, Yawson DO, Odoi JO, Afitiri AR, Nyieku FE (2018) Access to improved water and sanitation in sub-Saharan Africa in a quarter century. Heliyon 4(11):e00931

Beebeejaun Y (2017) Gender, urban space, and the right to everyday life. J Urban Aff 39(3):323–334

Bhanye J (2023) "Nimble sociality and belonging": an ethnography of migrants' responses to bans on associational life during the COVID-19 pandemic. In: Urban Forum. Springer Netherlands, Dordrecht, pp 1–27

Carrard N, Crawford J, Halcrow G, Rowland C, Willetts J (2013) A framework for exploring gender equality outcomes from WASH programmes. Waterlines 32(4):315–333. https://doi.org/10.3362/1756-3488.2013.033

Caruso BA, Conrad A, Salinger A, Patrick M, Youm A, Sinharoy S (2021) A conceptual framework to inform national and global monitoring of gender equality in WASH. WHO/UNICEF Joint Monitoring Programme for Water Supply, Sanitation, and Hygiene. Available at: https://washdata.org/sites/default/files/2021-10/jmp-2021-gender-review-conceptual-framework.pdf

Choudhary N, Brewis A, Wutich A, Udas PB (2020) Sub-optimal household water access is associated with greater risk of intimate partner violence against women: evidence from Nepal. J Water Health 18(4):579–594

Dugarova E (2015) Social inclusion, poverty eradication and the 2030 agenda for sustainable development (No. 2015–15). UNRISD Working Paper

Ferrant G (2015) How do gender inequalities hinder development? Cross-country evidence. Ann Econ Stat/Annales d'Économie et de Statistique 117/118:313–352

Fotio HK, Nguea SM (2022) Access to water and sanitation in Africa: does globalisation matter? Int Econ 170:79–91

Geere JAL, Hunter PR, Jagals P (2010) Domestic water carrying and its implications for health: a review and mixed methods pilot study in Limpopo Province, South Africa. Environ Health 9:1–13

Guzura T (2017) An overview of issues and concepts in gender mainstreaming. Afro Asian J Soc Sci 8(1):1–21

Hemson D (2016) Water, sanitation, and health: South Africa's remaining and existing issues. South Afr Health Rev 2016(1):25–34
Jalali R (2021) The role of water, sanitation, hygiene, and gender norms on women's health: a conceptual framework. Gender Perspect Int Dev 1(1):21–44
Kabeer N (2018) Gender, livelihood capabilities and women's economic empowerment: reviewing evidence over the life course. Gender and Adolescence: Global Evidence, London. Available at: https://eprints.lse.ac.uk/90462/
Kayser GL, Rao N, Jose R, Raj A (2019) Water, sanitation and hygiene: measuring gender equality and empowerment. Bull World Health Organ 97(6):438
Keddy KH, Smith AM, Sooka A, Tau NP, Ngomane HM, Radhakrishnan A, Als D, Benson FG (2018) The burden of typhoid fever in South Africa: the potential impact of selected interventions. Am J Trop Med Hyg 99(3 Suppl):55
Leach M (ed) (2015) Gender equality and sustainable development. Routledge
Leach M, Mehta L, Prabhakaran P (2016) Gender equality and sustainable development: a pathways approach. The UN Women Discussion Paper 13:2016
McCammon E, Bansal S, Hebert LE, Yan S, Menendez A, Gilliam M (2020) Exploring young women's menstruation-related challenges in Uttar Pradesh, India, using the socio-ecological framework. Sex Reprod Health Matters 28(1):1749342
Mocwagae K, Mphambukeli T (2023) Planning for effective and sustainable water access and provision in QwaQwa through the UN sustainable development goals. In: Sustainable futures in southern Africa's mountains: multiple perspectives on an Emerging City. Springer International Publishing, Cham, pp 105–126
Muller M, Schreiner B, Smith L, van Koppen B, Sally H, Aliber M, Cousins B, Tapela B, Van der Merwe-Botha M, Karar E, Pietersen K (2009) Water security in South Africa. Development Planning Division. Working Paper Series, 12
Mushavi RC, Burns BF, Kakuhikire B, Owembabazi M, Vořechovská D, McDonough AQ, Cooper-Vince CE, Baguma C, Rasmussen JD, Bangsberg DR, Tsai AC (2020) "When you have no water, it means you have no peace": a mixed-methods, whole-population study of water insecurity and depression in rural Uganda. Soc Sci Med 245:112561
National Planning Commission (NPC) (2012) National Development Plan 2030 https://www.gov.za/sites/default/files/gcis_document/201409/ndp-2030-our-future-make-it-workr.pdf
Nayar KR (1994) Politics of 'sustainable development'. Econ Polit Wkly:1327–1329
Ogunbanjo MA, Oyedokun DM, Oladipupo AO (2022) Global governance and sustainable development: a discussion from the socio-political and socio-ecological perspectives. Glob Gov 12(1):24–36
Prüss-Ustün A, Wolf J, Bartram J, Clasen T, Cumming O, Freeman MC, Johnston R (2019) Burden of disease from inadequate water, sanitation and hygiene for selected adverse health outcomes: an updated analysis with a focus on low-and middle-income countries. Int J Hyg Environ Health 222(5):765–777
Rampaul K, Magidimisha-Chipungu H (2022) Gender mainstreaming in the urban space to promote inclusive cities. J Transdiscipl Res S Afr 18(1):1–9
Republic of South Africa (RSA) (2018). https://www.gov.za/sites/default/files/gcis_document/201911/national-water-and-sanitation-master-plandf.pdf
Republic of South Africa Department of Water and Sanitation (RSA DWA) (2021/2022) Annual report (https://www.dws.gov.za/Documents/AnnualReports/DWS_2021-22AnnualReport.pdf)
Rossouw L, Ross H (2021) Understanding period poverty: socio-economic inequalities in menstrual hygiene management in eight low-and middle-income countries. Int J Environ Res Public Health 18(5):2571
Saleem M, Burdett T, Heaslip V (2019) Health and social impacts of open defecation on women: a systematic review. BMC PublicHealth 19(1):1–12
Sandercock L, Forsyth A (1992) Feminist theory and planning theory: The epistemological linkages. Planning Theory, (1992/7–8)
Scheihing KW, Tanner J, Weaver M, Schöniger M (2020) A strategy to enhance management of free basic water via communal taps in South Africa. Util Policy 64:101043
Scherer N, Mactaggart I, Huggett C, Pheng P, Rahman MU, Biran A, Wilbur J (2021) The inclusion of rights of people with disabilities and women and girls in water, sanitation, and hygiene policy documents and programs of Bangladesh and Cambodia: content analysis using EquiFrame. Int J Environ Res Public Health 18(10):5087
Shah V, Nabwera H, Sonko B, Bajo F, Faal F, Saidykhan M, Jallow Y, Keita O, Schmidt WP, Torondel B (2022) Effects of menstrual health and hygiene on school absenteeism and drop-out among adolescent girls in rural Gambia. Int J Environ Res Public Health 19(6):3337
Sharma R, Singh R (2012) Development of neural network model to classify women carrying water jobs in terms of physiological risk. Int J Comput Appl 975:8887
South African Government News Agency (SAnews.gov) (2023) Cholera death toll rises to 31 in SA, 166 laboratory-confirmed cases recorded. Available at https://www.sanews.gov.za/south-africa/cholera-death-toll-rises-31-sa-166-laboratory-confirmed-cases-recorded (Accessed 21 August 2023)
Statistics South Africa (StatsSA) General Household Survey (GHS) report (2018) (https://www.statssa.gov.za/publications/P0318/P03182018.pdf)
Statistics South Africa (StatsSA) General Household Survey (GHS) report (2021) (https://www.statssa.gov.za/publications/P0318/P03182021.pdf)
Tomberge VMJ, Bischof JS, Meierhofer R, Shrestha A, Inauen J (2021) The physical burden of water carrying and Women's psychosocial Well-being: evidence

from rural Nepal. Int J Environ Res Public Health 18(15):7908
Troeger C, Khalil IA, Rao PC, Cao S, Blacker BF, Ahmed T, Armah G, Bines JE, Brewer TG, Colombara DV, Kang G (2018) Rotavirus vaccination and the global burden of rotavirus diarrhea among children younger than 5 years. JAMA Pediatr 172(10):958–965
UNDP Water Governance Facility/UNICEF (2015) "WASH and accountability: explaining the concept" accountability for sustainability partnership: UNDP water governance facility at SIWI and UNICEF. Stockholm and New York. Available from http://www.watergovernance.org/
United Nations (UN) (2023) The Sustainable Development Goals Report (https://unstats.un.org/sdgs/report/2023/The-Sustainable-Development-Goals-Report-2023.pdf)
United Nations Children's Fund (UNICEF), World Health Organization (WHO) (2020) State of the world's sanitation: An urgent call to transform sanitation for better health, environments, economics and societies. WHO
United Nations Children's Fund (UNICEF), World Health Organization (WHO) (2023). https://washdata.org/reports/jmp-2023-wash-households
Venkataramanan V, Geere JAL, Thomae B, Stoler J, Hunter PR, Young SL (2020) In pursuit of 'safe' water: the burden of personal injury from water fetching in 21 low-income and middle-income countries. BMJ Glob Health 5(10):e003328
Wardeh M, Marques RC (2023) Progress on sustainable development goal 6 in refugee camps in the Middle East: a comparative study. Util Policy 82:101575
WHO/UNICEF Joint Monitoring Programme (JMP) (2023) for Water supply, sanitation, and hygiene. https://www.unwater.org/publications/who/unicef-joint-monitoring-program-update-report-2023
Witherington B III, Witherington B (1987) Women in the ministry of Jesus: a study of Jesus' attitudes to women and their roles as reflected in his earthly life (no. 51). Cambridge University Press
World Health Organization (WHO) (2019). https://apps.who.int/iris/bitstream/handle/10665/330100/WHO-CED-PHE-WSH-19.149-eng.pdf
World Health Organization (WHO) (2021) Progress on household drinking water, sanitation and hygiene 2000–2020: five years into the SDGs
Wurster A (2020) Ready for water-shedding? Water&Sanitation Africa 15(1):8–9
Young IM (2005) On female body experience: "throwing like a girl" and other essays. Oxford University Press
Young IM (2008) Structural injustice and the politics of difference. In: Intersectionality and beyond. Routledge-Cavendish, pp 289–314
Zerbo A, Delgado RC, González PA (2021) Conceptual frameworks regarding waterborne diseases in sub-Saharan Africa and the need of for a new approach to urban exposomes. Epidemiol Health 43

Building Resilient Urban Infrastructure and Innovation for Urban Public Health: A Case Study of Bloemfontein, South Africa

5

Siphokazi Rammile, Johannes I. Bhanye, Timothy Lehobo, and Mareli Hugo

5.1 Introduction

The dynamic shift towards urbanisation is reshaping the world's demographic landscape, with more than half the population residing in urban areas. This trend is projected to escalate, with urban populations comprising 68% of the world's inhabitants by 2050 (Cohen 2015). The UN World Population Prospects predicts that over half of the forthcoming global population growth will be in Africa (United Nations DESA 2017). Thus, the discussion around the need for improved, innovative urban infrastructure in Africa is very relevant. When numerous African cities were developed, they were not equipped for the vast populations they now house. Population increases and land use changes placed immense pressure on urban infrastructure (Abuya 2020). Considering the increased pressures and insufficient maintenance and innovation, it is no wonder that the aesthetics of many South African cities are overflowing sewer systems and burst service pipes. Amid the urban surge, the obligation to address complex challenges intertwined with urbanisation becomes more pressing than ever, particularly in the realm of urban public health.

African cities that were once hubs of innovation, culture, and economic prosperity are now confronting a range of urban public health threats emanating from their dense urban form. The convergence of infectious diseases, non-communicable ailments, injuries, and the unprecedented impact of global events such as the recent COVID-19 pandemic revealed the urgency of fortifying urban public health infrastructure (Matamanda et al. 2022; Gatzweiler et al. 2021; Biglieri et al. 2021). In this context, urban development grounded in Sustainable Development Goals (SDGs) is critical for safeguarding urban public health and fostering innovation within urban environments.

The SDGs were established to build upon the Millennial Development Goals (MDG) and compensate for the shortfalls, with some goals directly linked to the MDG and others addressing the gaps (Puplampu et al. 2017). There are six foci in the 11 SDGs, namely dignity, people, planet, partnership, justice and prosperity (UN 2015). The overall goal of SGD 9 is directly linked to prosperity to promote sustainable

S. Rammile (✉) · T. Lehobo · M. Hugo
Department of Urban and Regional Planning, University of the Free State, Bloemfontein, South Africa

J. I. Bhanye
Department of Urban and Regional Planning, University of the Free State, Bloemfontein, South Africa

African Climate and Development Initiative (ACDI), University of Cape Town, Cape Town, South Africa

A. R. Matamanda, V. Nel (eds.), *Sustainable Development Goals and Urban Health*, Sustainable Development Goals Series, https://doi.org/10.1007/978-3-031-68734-1_5

industry, develop resilient infrastructure, and nurture innovation. Gatzweiler et al. (2021) emphasise the need for community buy-in to ensure the longevity and resilience of the urban fabric, as urban residents need to assume responsibility for developments in their neighbourhoods. This is why it was essential for this study to engage with community members in marginalised communities to give them a voice.

This chapter examines the relationship between SDG 9 and urban public health, reflecting on successes, pitfalls and inherent gaps in attaining this goal and how they affect urban public health. The study argues that a focus on building resilient infrastructure and fostering innovation for urban public health in the African city's marginalised communities is urgently required to support the SDGs and other global and regional public health targets. Data for the study were based on secondary sources triangulated with key informant interviews and in-depth interviews. We explored the following research questions:

- What are the current urban public health challenges in the selected marginalised communities?
- What is the state of the existing urban public health infrastructure and innovation in the marginalised communities?
- What are the existing institutions responsible for urban public health infrastructure and innovation in marginalised communities?
- What are the new forms of infrastructure and innovation that can be put in place to improve the state of urban public health in the marginalised communities identified?

The chapter initially explores the conceptual and theoretical foundations of SDG 9 and urban public health through the lens of resilience, considering the specific targets and indicators aligned with the SDG. The relationship between urban public health, infrastructure and innovation is explored globally before considering the South African context of SDG 9. The targets and indicators set for the SDG at a national level are investigated, along with the progress made in its attainment. The chapter analyses the challenges and vulnerabilities of urban infrastructure in the country, examining how this affects urban public health. The COVID-19 pandemic, in particular, has exposed the need for resilient urban public health infrastructure and innovation in African cities, which is explored in this chapter. The results of the primary data collected in four Mangaung townships are analysed, feeding into the policy alignment and closing discussion of the chapter.

5.2 Overview of Sustainable Development Goal 9 and Urban Public Health

Long-term economic development requires innovation, inclusive and sustainable industries, and well-maintained physical infrastructure (United Nations 2020). With advancements in technology and connectivity, governments, businesses, and industries can effectively address the challenges faced by society. Many countries lack basic resilient infrastructure, hindering industries, businesses, and communities to thrive (United Nations 2020:44; Gaal and Afrah 2017). SDG 9 focuses on infrastructure development and sustainable industries to build resilient infrastructure, promote inclusive and sustainable industrialisation, and foster innovation, directly affecting urban public health (Tomaselli et al. 2019).

Urban public health is a multidisciplinary field that promotes and protects the health and well-being of individuals and communities within urban areas. Health does not merely refer to the absence of injury or disease but rather a state of physical, mental and social well-being (WHO Constitution 1989). Well-being directly relates to a positive state of being, which can be aligned with various dimensions; Husain and Akram (2022) narrow them down to the quality of life (subjective) and standard of living (objective). To be effective, the indicators and dimensions of well-being must be subjective and objective, taking cognisance of diverse communities and variations in perceptions.

Urban public health recognises the diverse and dynamic nature of urban populations (D'Alessandro 2020). It involves assessing and monitoring the health status of urban residents, identifying health disparities, and implementing interventions to improve overall population health (Capolongo et al. 2020). This includes understanding urban communities' demographic composition, health behaviours, and health outcomes. Urban public health is greatly influenced by income, education, employment, social support, and community cohesion within urban settings. Conceptualising urban public health involves addressing social inequities and creating supportive environments that enable individuals and communities to lead healthy lives. Infrastructure and transportation can increase accessibility to essential services such as health care, education, and employment. Streetlights and well-maintained sidewalks can reduce the incidence of crime and violence, making people feel safer and more secure. Open green spaces, pedestrian paths and bike paths can improve a community's environmental quality, positively influencing health and well-being (Lennon et al. 2017). Various targets and indicators are set to monitor the progress of industries working towards achieving SDG 9 (Kynčlová et al. 2020). With increased investment in innovation and infrastructure development, it is possible to bridge the digital divide, promote sustainable industrial development, and invest in scientific research and innovation (UNDP 2015).

5.2.1 The Role of Resilient Urban Infrastructure and Innovation in Urban Public Health

Designing, constructing, and managing infrastructure systems capable of withstanding and recovering from shocks, stresses, and disturbances in cities are essential to resilient urban infrastructure (Francis and Bekera 2014). Using durable materials and engineering practices, resilient urban infrastructure can withstand shocks and stresses (Hugo and Bhanye 2022). It maintains structural integrity, allowing essential services to continue during and after events like earthquakes, floods, or disease outbreaks. Redundancy infrastructure, initially used in information technology and commonly used in electrical grid systems, can be implemented in the built environment to increase the resilience of urban infrastructure. Incorporating redundancy serves as a backup mechanism, preserving service continuity even if one component fails or is damaged (Forcellini 2023). The aim of incorporating redundancy is to ensure that provision is made for more than what is needed. Road networks should connect urban areas so that when one road is closed, people are still connected to economic activities, employment, healthcare and education through alternative routes. Redundancy is the repetition of existing infrastructure to allow alternative options if the primary infrastructure fails. By neutralising the effects of disruptions, resilient infrastructure sustains functionality during critical situations. Addressing vulnerabilities and implementing protective features minimise the potential for damage and loss during disruptive events, enhancing the resilience of the infrastructure.

Resilient urban infrastructure is also characterised by flexibility and adaptability to changing circumstances. It accommodates evolving needs and anticipated challenges, including technological advancements, population growth, climate variations, and alterations in land use (Chester and Allenby 2019). The ability to adjust and modify infrastructure enables proactive responses to emerging challenges and utilising new opportunities. Engaging with communities and considering their needs, perspectives, and vulnerabilities are integral components of building a resilient urban infrastructure. Social considerations, such as equity, inclusivity, and accessibility, ensure that infrastructure benefits all residents, including marginalised groups (Bhanye and Dzingirai 2020; Bhanye and Dzingirai 2022). Through community engagement, resilience-building efforts become more effective and inclusive, fostering ownership, awareness, and co-operation (Meerow et al. 2019).

Resilient urban infrastructure plays a crucial role in urban public health and achieving SDG 9.

One aspect of resilient urban infrastructure is providing adequate and reliable access to essential services such as clean water, sanitation, and healthcare facilities. Sustainable infrastructure development should prioritise the provision of safe drinking water and improved sanitation facilities for urban populations, reducing the risk of disease outbreaks and improving overall urban public health (Carbonell et al. 2023). Furthermore, resilient infrastructure includes well-designed and accessible transportation systems, including roads, public transportation, and pedestrian and cycling networks (Yannis and Chaziris 2022). Efficient transportation systems can contribute to improved urban public health by reducing air pollution and promoting physical activity through active transportation options. Sustainable urban planning and infrastructure development should prioritise the creation of walkable and bike-friendly cities, along with reliable and affordable public transportation, to enhance urban public health (Pozoukidou and Chatziyiannaki 2021).

Resilient urban infrastructure is closely linked to sustainable urbanisation and creating inclusive and safe cities. SDG 11, as discussed in this book, focuses on making cities and human settlements inclusive, safe, resilient, and sustainable, interlinked with SDG 9 (Franco et al. 2019). Resilient infrastructure ensures that cities can effectively respond to and recover from shocks and stresses, such as natural disasters or urban public health emergencies (Corvalan et al. 2020). During the COVID-19 pandemic, marginalised communities struggled to abide by the national safety precautions as they had limited access to COVID-19 testing, could not quarantine properly, and had limited access to clean water for hand washing and healthy meals. Resilient urban infrastructure is critical for these communities and provides access to needed services during disasters. Furthermore, investing in resilient urban infrastructure can contribute to several other SDGs, such as SDG 7 (Affordable and Clean Energy) by promoting renewable energy sources, SDG 11 (Sustainable Cities and Communities) by creating sustainable and liveable urban environments, and SDG 13 (Climate Action) by reducing vulnerability to climate change impacts (Franco et al. 2019).

Resilient urban infrastructure is intertwined with urban public health and the achievement of SDG 9. Resilient infrastructure can improve urban public health outcomes by prioritising access to essential services, promoting sustainable transportation options, and addressing climate change impacts. The targets and indicators aligned with SDG 9 provide a framework for monitoring progress and guiding efforts to improve urban public health as part of sustainable development initiatives. Effective monitoring and implementation of targets and indicators related to urban public health and resilient infrastructure are essential to develop healthier and more resilient communities.

5.2.2 Assessment of SDG 9, Urban Public Health, Infrastructure and Innovation

Specific targets and indicators can be used to ensure progress towards resilient urban infrastructure and its impact on urban public health. The United Nations defined eight targets along with 12 indicators for SDG 9 (NU CEPAL 2019:43). The targets specify the goals, while the indicators represent the metrics to track the extent to which the targets have been achieved (See Fig. 5.1). While all eight targets are equally essential to achieving the goal, not all speak to urban public health.

5.2.2.1 Infrastructure, SDG 9 and Urban Public Health

The built environment plays a role in urban public health. It encompasses physical infrastructure, transportation systems, housing conditions, green spaces, and access to essential services within urban areas. Conceptualising urban public health involves examining how the built environment influences health outcomes and designing urban spaces that promote physical activity, safety, and access to healthcare services. Target 9.1 focuses on developing quality, reliable, sustainable, and resilient infrastructure (Tomaselli et al. 2019). Developing trans-border infrastructure will support economic growth, community well-being

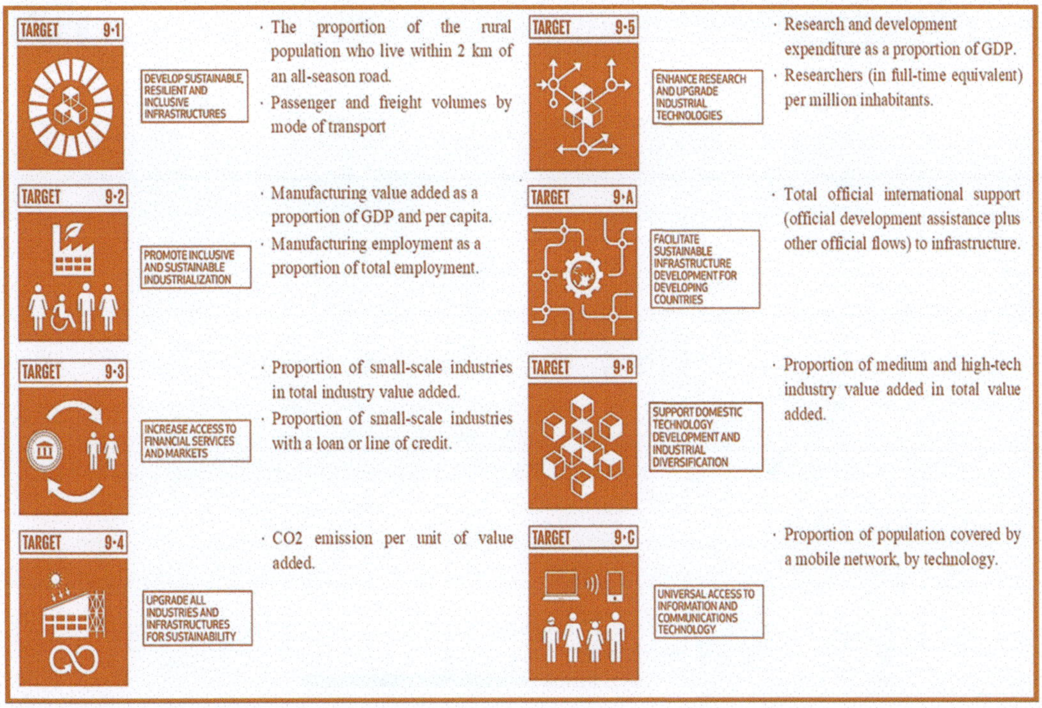

Fig. 5.1 SDG 9 targets and indicators. (Source: Author)

and urban public health. The target is to promote affordable and equitable access for all (NU CEPAL 2019). Indicators for Target 9.1 include available modes of transport, transport infrastructure and distance from transportation hubs within rural areas (United Nations 2020). Figure 5.2 indicates the percentage of the rural population that lives within the set 2 km from transportation hubs and networks as of 2021.

There is limited data for this target, and the available data indicates that rural areas do not have access to adequate, connected transportation networks. This target emphasises the importance of infrastructure that is not only built to last but is also designed with sustainability and resilience in mind. It involves improving access to reliable and affordable transportation systems, energy services, and information and communication technologies (OECD 2018:9). Improving infrastructure and connectivity can positively affect the local economy and urban public health. Not only does this speak to road infrastructure, but air transportation as well (Fig. 5.3).

Target 9.1 (passenger and freight volumes) can play a critical role in the trade of landlocked countries and provide necessary medical aid and resources in times of crises/ natural hazards, ensuring these countries can move towards resilient development (Adewole and Struthers 2018). Urban public health is affected by connectivity and access to health care and medical supplies. In a state of disaster, marginalised communities rely on support from external sources; therefore, it is essential to have resilient transport infrastructure in place. As reflected in Fig. 5.3, the freight ton-km for developing countries is especially low, which indicates limited connectivity in these regions.

5.2.2.2 Small Businesses, SDG 9 and Urban Public Health

The third target (Target 9.3) speaks to the development and investment within small-scale industrial enterprises. This development is possible through financial support, affordable credit and the integration of small-scale businesses into

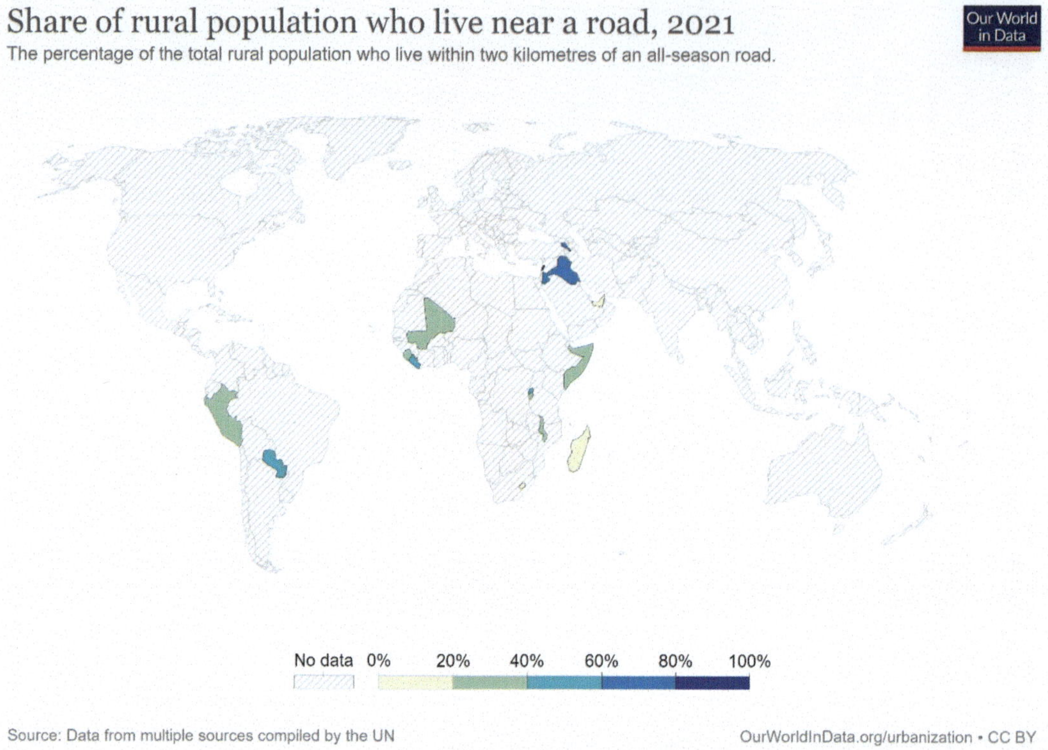

(Source: Goal 9: Industry, innovation and infrastructure - SDG tracker, 2020:online.)

Fig. 5.2 Share of the rural population who live near a connected road as of 2021. (Source: Goal 9: Industry, innovation and infrastructure—SDG tracker 2020:online)

value chains and markets (Tomaselli et al. 2019). The value added to the Gross Domestic Product (GDP) and the proportion of small-scale industries indicate the extent to which the target has been reached. Small-scale enterprises can improve local economic development as they are a source of employment in developing countries and emerging economies (United Nations 2020). These small entrepreneurs are more vulnerable and do not have the capacity to deal with unexpected shocks, affecting the state of urban public health. During the COVID-19 pandemic, many small entrepreneurs had to close down. Providing financial services to small- and medium-sized enterprises is vital to ensure they survive and thrive during and after crises (United Nations 2020). Access to financial support can increase the competitiveness of these entrepreneurs, enabling them to form part of local and global markets and overall increase the country's economic growth and infrastructure development, thus improving the well-being of communities (OECD 2017).

Transportation infrastructure development, industries and small entrepreneurs faced challenges during the COVID-19 pandemic, which led to disruptions in the global value chain and employment stability. In developing countries, manufacturing jobs are an essential source of income and are crucial to poverty reduction and the well-being of communities and urban areas. The after-effects of COVID-19 have threatened to halt or even reverse progress towards SDG 9 and related goals.

5.2.2.3 Pollution, SDG 9 and Urban Public Health

Urban areas face unique environmental health challenges due to population density, industrial activities, and exposure to pollutants.

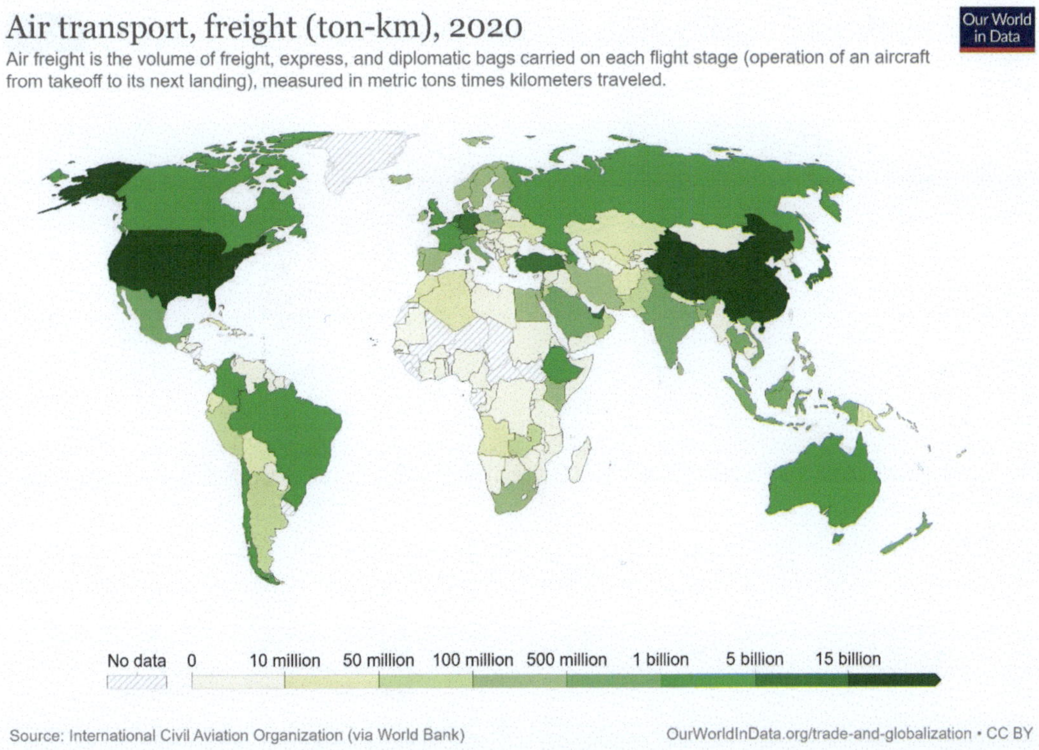

(Source: Goal 9: Industry, innovation and infrastructure - SDG tracker, 2020:online.)

Fig. 5.3 Air transport in freight for 2020. (Source: Goal 9: Industry, innovation and infrastructure—SDG tracker 2020:online)

Conceptualising urban public health includes monitoring air quality, water quality, and waste management systems to ensure a safe and healthy environment. It also involves addressing environmental justice issues and promoting sustainable practices to mitigate the adverse health effects of urban living.

The health of the environment also influences urban public health. The need for production can negatively impact the health of the environment because of increased pollution. Sustainable industries and efficient use of resources, with the help of technological advancements, can address pollution and CO_2 emissions from industries (Target 9.4) (NU CEPAL 2019). This is critical, as many communities in rural areas and mining areas face health and environmental challenges due to unsustainable industrial development and mining activities.

As indicated in Fig. 5.4, developing countries display a higher carbon emission intensity than developed countries, and the carbon emission footprint in South Africa is very high (compared to the size of the economy). This negatively affects the natural environment and communities, as the high prevalence of carbon emissions correlates to poor air quality. Poor air quality is disastrous for those who do not have access to personal protective measures, particularly the most vulnerable.

5.2.2.4 Technology, SDG 9 and Urban Public Health

The need for technological development in African countries is emphasised with Target 9 A. The target aims to facilitate sustainable and resilient infrastructure development through enhanced financial support to least-developed

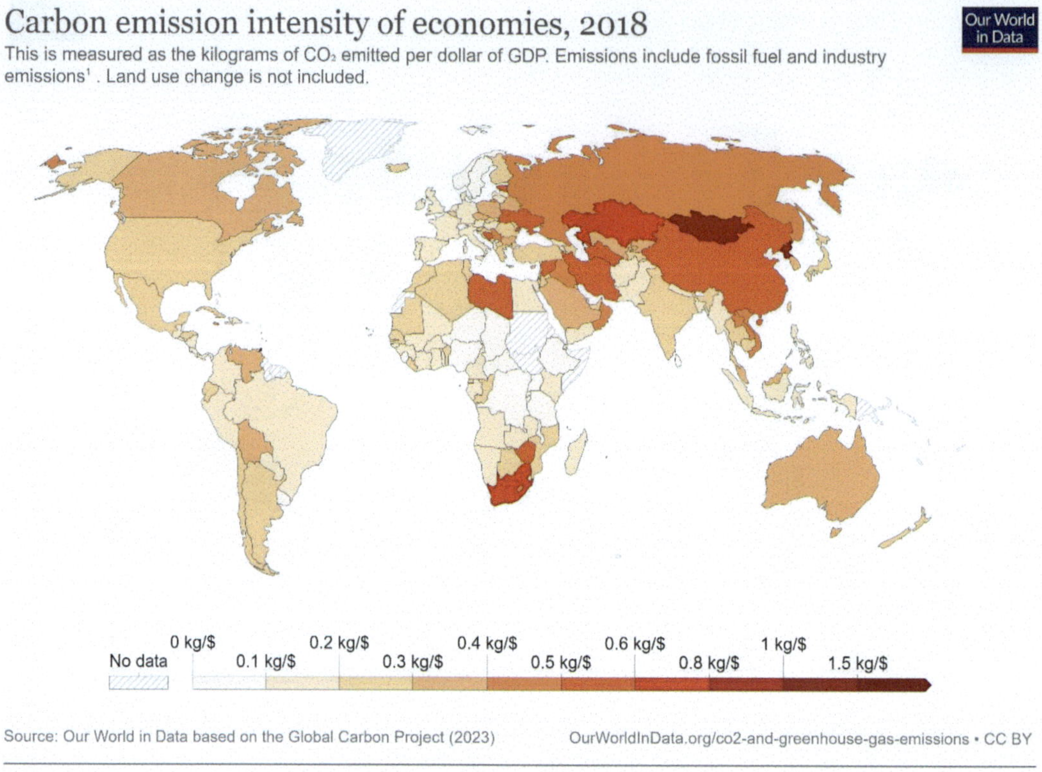

(Source: Goal 9: Industry, innovation and infrastructure - SDG tracker, 2020: online.)

Fig. 5.4 Carbon emission intensity of economies for 2018. (Source: Goal 9: Industry, innovation and infrastructure—SDG tracker 2020: online)

countries, landlocked developing countries and small islands (Luken et al. 2022). Indicators include official development assistance plus other official flows on an international level and total investment in infrastructure development. Target 9 B addresses the need for support with technological development and innovation within developing countries (ECLAC 2021). A beneficial policy environment can further promote industrial diversification and add value to goods and services. To ensure connectivity between countries, access to information and communications technology should be addressed (Target 9 C) (OECD 2018). Developing countries can achieve this by providing affordable internet access and increasing coverage to reach more people. During 2021, 88% of the world population had 4G internet access; progress has been made in mobile connectivity (Fig. 5.5). With the world moving into the fourth industrial revolution, it is essential for these countries to invest in technological infrastructure and improve internet access.

5.2.3 Global Summary of SDG 9 and Urban Public Health

Industrial development ensures an increase in manufacturing goods and providing employment to communities. Innovation, inclusive and sustainable industries, and well-maintained physical infrastructure are essential for long-term economic development. However, many countries face challenges due to inadequate resilient infrastructure, hindering the growth of industries, businesses, and communities. SDG 9 focuses on infrastructure development and sustainable industries, directly affecting urban public health. Resilient urban infrastructure is designed to with-

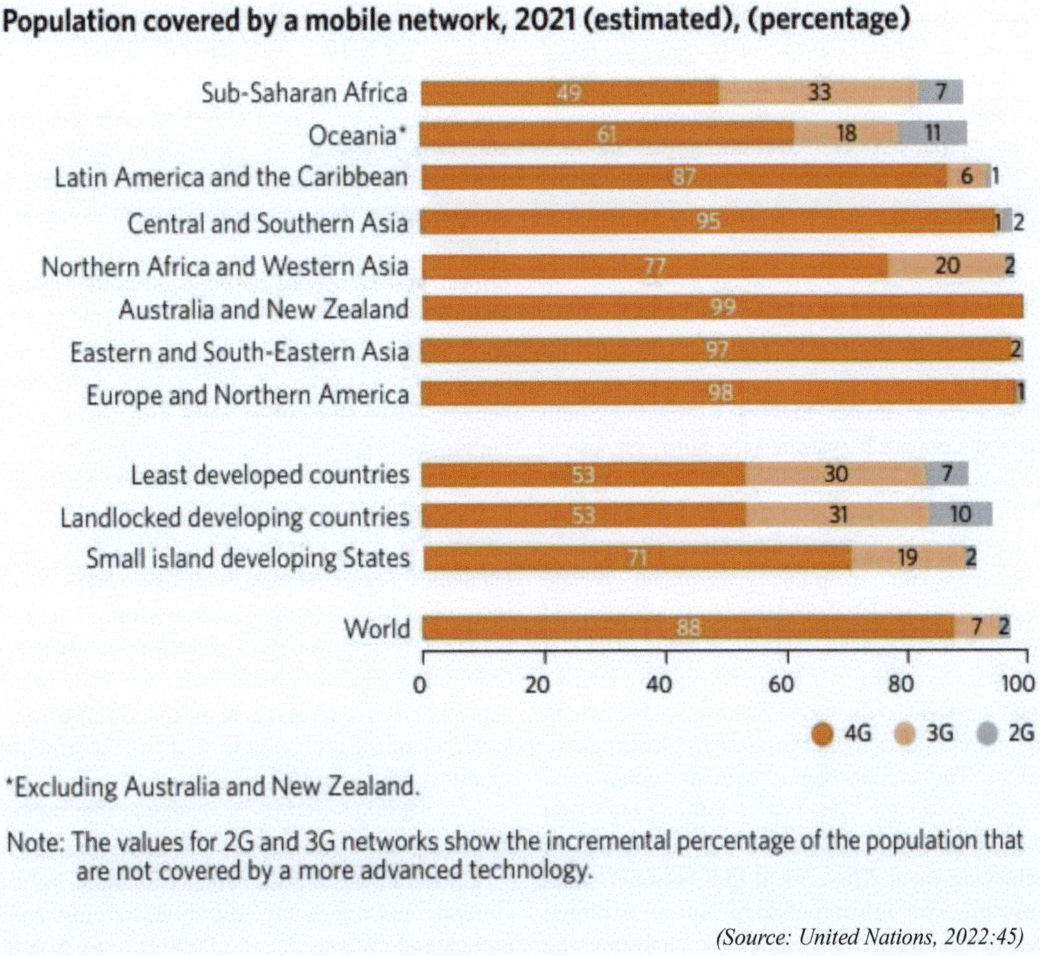

Fig. 5.5 Estimated percentage of the world population that has internet access. (Source: United Nations 2022:45)

stand shocks and stresses, ensuring the continuity of essential services. It incorporates redundancy, flexibility, and adaptability to address changing circumstances.

Measures are taken to mitigate risks and protect against potential hazards. Integration and interconnectivity among infrastructure systems enhance coordination and response during disruptions. Sustainable practices and community engagement promote inclusivity and minimise environmental impact. Resilient urban infrastructure contributes to improved urban public health outcomes by providing access to essential services, promoting sustainable transportation options, and enhancing overall resilience. Achieving SDG 9 requires monitoring progress and implementing initiatives that support sustainable development, connectivity, and small-scale industrial development.

The COVID-19 pandemic has highlighted the importance of resilient infrastructure, particularly in supporting small entrepreneurs and ensuring the continuity of industries. Addressing environmental health challenges and promoting sustainable practices are integral to conceptualising urban public health. Technological development and enhanced financial support to developing countries are crucial for sustainable and resilient urban infrastructure development. Access to information and communication technology, along with internet connectivity, is vital for technological advancement and industrial diversification. Improving infrastructure and connectivity will contribute to economic growth, enhance the well-being of communities, and mitigate the adverse effects of unsustainable industrial activities.

5.3 Context of SDG 9 in South Africa

This section situates SDG 9 within the South African context. It examines the aims, successes, and shortcomings in achieving SDG 9 in South Africa, particularly concerning urban public health. To achieve the SDGs, stakeholder collaboration is vital, including government, private sector, civil society, academic institutions, and international agencies (Haywood et al. 2019). It is crucial to evaluate the progress made in achieving the SDGs so that areas needing improvement are identified and addressed (Wernecke et al. 2021). The section begins with the targets and indicators set for SDG 9 at a national level and then proceeds to assess the progress made in achieving them. The critical link between infrastructure and urban public health is explored before the section concludes by analysing the current trends and technologies to improve urban public health infrastructure and their applicability in South Africa. Urban public health is a multidisciplinary concept encompassing mental, physical and community health. To achieve resilient urban public health, stakeholder collaboration and innovation are critical. Olszewska-Guizzo et al. (2022) establish that the key stakeholders necessary for maintaining urban public health are:

- Medical professionals
- Non-governmental organisations
- Design professionals
- Researchers
- Educators
- Government
- Communities

5.3.1 Targets and Indicators Set for SDG 9 at National Level

The South African Constitution states the inalienable right to life for all, the right to health care and support services, as well as the right to an environment that poses no health threats (Republic of South Africa 1996). These rights inform and direct all subsequent policies and actions towards urban public health in the country. It is a mandate that deserves attention from the government and all other stakeholders alike because addressing public health directly impacts human well-being.

South Africa aims to ensure access to vital services across the country, to enhance equality and citizens' ability to attain economic improvement (Statistics South Africa 2019). The NDP then recognises the need to overcome infrastructure backlogs to improve the national health system (National Planning Commission 2012). It indicates the need to maintain and expand infrastructure to enhance economic and social growth (National Planning Commission 2012).

Against this backdrop, it is evident that South Africa has strong targets towards achieving SDG 9. These include consolidating economic infrastructure and improving education, training, and innovation (Wernecke et al. 2021). In South Africa, tracking progress towards attaining the SDG targets is done nationally to compare against global scales and progress (Wernecke et al. 2021). Statistics South Africa (2019) indicates that the specific SDG 9 targets South Africa aims to achieve are:

- 9.1—Implement high-quality economic regulation of network industries
- 9.2—Promote higher quality and industry-relevant education and training
- 9. 3—Encourage innovation and entrepreneurship

South Africa focused on eight SGD 9 targets and was able to report on seven indicators (Statistics South Africa 2019), as the following subsection demonstrates.

5.3.2 Progress Made Towards SDG 9 in South Africa

Since the end of apartheid in 1994, South Africans have experienced increased access to essential services and infrastructure, although inequalities caused by apartheid persist (Statistics South Africa 2019). The country has progressed on seven SDG 9 indicators (Statistics South Africa 2019) and has made strides in alleviating extreme poverty in some forms, including healthcare access (Fofana et al. 2018). However, significant challenges remain (Sustainable Development Solutions Network 2022); Table 5.1 shows South Africa's moderate progress towards achieving the SDGs.

Table 5.2 shows the specific performance indicators of South Africa's progress towards achieving SDG 9. Among others, it shows that access to information and communications technology (ICT) is increasing, transport infrastructure is stagnant, and research output is rising. All these have implications for public urban public health infrastructure and access.

The scientific and academic sectors play a crucial role in achieving SDG 9 as they lead training, skills development, research, and innovation, aligning with SDG 9 (Haywood et al. 2019). The South African Government, in collaboration with the UN in South Africa, developed the United Nations Sustainable Development Co-operation Framework (UNSDCF) for 2020 through 2025 (United Nations in South Africa 2023) to support the implementation of the SGDs in the country. If the potential of this tool is maximised, strides can be made in realising targets and being on track to achieve the SDG.

Table 5.1 South Africa's performance towards achieving the 17 SDGs

SDG	9			
Level of achievement				
Decreasing	Stagnant/ no change	Moderately increasing	On track	Status unknown N/A

Source: Adapted from UN SDG Data

Implementation of SDG 9 has seen a moderate increase. Table 5.3 summarises South Africa's progress towards achieving SDG 9. Among other indicators, the data shows an increase in scientific research, which is a positive factor for innovation. It also indicates an increase in the proportion of the population with access to ICT services. Research, innovation, and improved infrastructure help enhance urban public health in South Africa.

5.3.3 How SDG 9 Affects Urban Public Health in South Africa

One of the targets of SDG 9 is to: *"facilitate sustainable and resilient infrastructure development in developing countries through enhanced financial, technological and technical support to African countries, least developed countries, landlocked developing countries and small island developing countries"* (Ramirez-Rubio et al. 2019: 8). Demonstrating the need to equip countries of the global south with adequate infrastructure to achieve social, economic, and spatial development. Urban public health is a vital element of development, which relies on adequate infrastructure.

In South Africa, the focal SDG 9 elements are improving access to road infrastructure, education and training, and innovation in ICT (Statistics South Africa 2019). Adequate and resilient infrastructure, including transportation, water and sanitation, and medical facilities and technologies, is critical to improving urban public health access for the urban population in South Africa. This means that achieving SDG 9, specifically the aforementioned targets and indicators, will positively impact urban public health in the country. Therefore, a crucial relationship and interlink exists between urban public health and infrastructure in South Africa.

Table 5.2 South Africa's SDG 9 performance by indicator

South Africa SDG 9 performance by indicator			
SDG 9: industry, innovation, and infrastructure	Value	Year	Trend
Rural population with access to all-season roads (%)	94.6	2022	–
Population using the internet (%)	72.3	2021	↑
Mobile broadband subscriptions (per 100 population)	115.7	2021	↑
Logistics performance index: Quality of trade and transport-related infrastructure (worst 1–5 best)	3.2	2018	→
The times higher education universities ranking: Average score of top 3 universities (worst 0–100 best)	51.1	2022	–
Articles published in academic journals (per 1000 population)	0.5	2021	↑
Expenditure on research and development (% of GDP)	0.6	2019	↓

Source: Adapted from UN SDG Data

5.3.4 Challenges and Vulnerabilities of Urban Infrastructure in South Africa

South Africa recognises the importance of infrastructure for national economic growth, especially for uplifting previously marginalised communities out of poverty. However, this has not translated into any national policy for sustainable urban infrastructure (Palmer et al. 2016). Although there has been increased access to essential infrastructure services in urban areas, there remains a shortage in urban informal settlements and peri-urban areas (National Planning Commission 2012). This disparity highlights a notable vulnerability in South African urban infrastructure, where the most vulnerable groups remain under-served.

Abrams et al. (2021) show that there is inadequate access to water, sanitation, and hygiene (WASH) services in the country, especially in remote rural areas and small towns, where a lack of skills, capacity, and funding hinders equitable access. There remains room for improvement in the state of urban infrastructure in South Africa if adequate urban public health is to be realised.

5.3.5 The Link Between Infrastructure and Urban Public Health

Infrastructure and urban public health are intricately linked. Abrams et al. (2021) indicate that access to WASH services is vital for urban public health. Zimmerman (2005) further demonstrates that mass transit infrastructure is essential for access to urban public health and is used globally to good effect. Benefits of mass transit infrastructure include lowering toxic gas emissions, improving air quality, saving energy use, and enhancing safety due to slower speeds, all of which contribute to human health and well-being.

Azzopardi-Muscat et al. (2020) note that the built environment impacts urban public health and that architects tend to allocate limited space to health. The authors further indicate that planners and architects have a critical role in urban public health, considering its evolution from a purely bio-medical issue to a more socio-anthropological and spatial concern. Therefore, the planning, designing, and location of all built-up structures must consider the likely impacts on the environment and the health and well-being of human beings. The right to health includes essential infrastructure services, including water, sanitation, and education—without which proper health cannot be attained (Van de Pas et al. 2017). Water aids in cleaning, bathing, cooking, and drinking; sanitation aids in the hygienic disposal of waste; and health education equips the populace with the appropriate information to make informed decisions about their health and lives. Without all these, there remains a shortage in adequate urban public health.

Access to quality internet connection affects urban public health because it aids in accessing digital services and information critical for health (Schram et al. 2018). Schram et al. (2018) examined the implementation of the National Broadband Network (NBN) in Australia, including the rollout of digital infrastructure to areas of different socioeconomic classes. They found that areas of lower socioeconomic status were more likely to receive lower-quality infrastructure, thus worsening health inequalities.

Table 5.3 South Africa's SDG 9 progress indicator

Indicator	Key data points
9.1.2D: Passenger and freight volumes by mode of transport (domesticated indicator)	Rail passengers 558 m (2008), 475 m (2015), 373 m (2016), 366 m (2017)
	Road passengers 287 m (2008), 330 m (2015), 283 m (2016), 281 m (2017)
	Rail freight 182 m tons (2008), 224 m tons (2015), 200 m tons (2016), 203 m tons (2017)
	Road freight 556 m tons (2008), 595 m tons (2015), 567 m tons (2016), 574 m tons (2017)
9.2.1: Manufacturing value added as a proportion of GDP and per capita	Manufacture value-added as a proportion of GDP 14% (2008), 12.8% (2013),12.4% (2015), 12.5% (2016), 12.3% (2017)
9.2.2: Manufacturing employment as a proportion of total employment	14.3% (2008), 12.2% (2013), 11.2% (2015), 10.7% (2016), 11% (2017)
9.5.1: Research and development expenditure as a proportion of GDP	0.89% (2008), 0.72% (2013), 0.8% (2015)
9.5.1A: Business expenditure on R&D (BERD) as a percentage of gross domestic expenditure on R&D (GERD)	57.7% (2007), 47.1% (2011), 45.9% (2013), 42.7% (2015)
9.5.2: Researchers (in full-time equivalent) per million inhabitants	398.6 (2007), 369 (2010), 438.2 (2013), 475.1 (2015)
9.b.1: Proportion of medium and high-tech industry value added in total value added	25.54% (2007), 24.43% (2011), 24.43% (2015)
9. c.1: Percentage of population covered by a mobile network, by 3G and LTE	3G: 97.8% (2015), 99.3% (2017), 99.5% (2018)
	LTE: 53% (2015), 76.7% (2017), 85.7% (2018)

Source: Statistics South Africa, 2019

It is thus inferred that the most vulnerable groups ought to be prioritised in infrastructure provision for adequate and equitable urban public health to come to fruition. Figure 5.6 demonstrates the infrastructure dynamics in rural, peri-urban, and urban areas. Rural areas tend to have abundant natural and social infrastructure while lacking the physical and institutional infrastructure necessary to thrive. Peri-urban areas have a balance in the types of infrastructure; however, they are characterised by low access levels. In contrast, urban areas often lack natural and social infrastructure, which are instrumental for urban public health.

5.3.6 Emerging Trends and Technologies

The concept of Smart Cities provides fertile ground for using technology in urban public health. Smart City professionals can employ strategies to enhance urban public health and safety. These include using thermal cameras or Internet of Things (IoT) sensors for early detection and management of disaster outbreaks (Allam and Jones 2020). These authors further indicate that incorporating artificial intelligence (AI) into IoT technologies can augment urban public health benefits.

These technologies help detect temperature spikes in human beings, which are critical in screening and monitoring. The technologies were used to good effect early after the outbreak of COVID-19, specifically in China. Wu et al. (2021) show that internet hospitals improved healthcare access and equity during the pandemic, reliable information platforms curtailed misinformation, and mobile health technologies bolstered China's fight against the outbreak. These may apply in South Africa, a reasonably advanced middle-income country with good technological capabilities.

Health in All Policies (HiAP) is an approach that methodically considers all the health implications of public policy to sidestep harmful health impacts of public policies and improve citizen health (Ramirez-Rubio et al. 2019). The HiAP approach facilitates health interventions, which aid in attaining SDG targets (Ramirez-Rubio et al. 2019). The authors assert that imple-

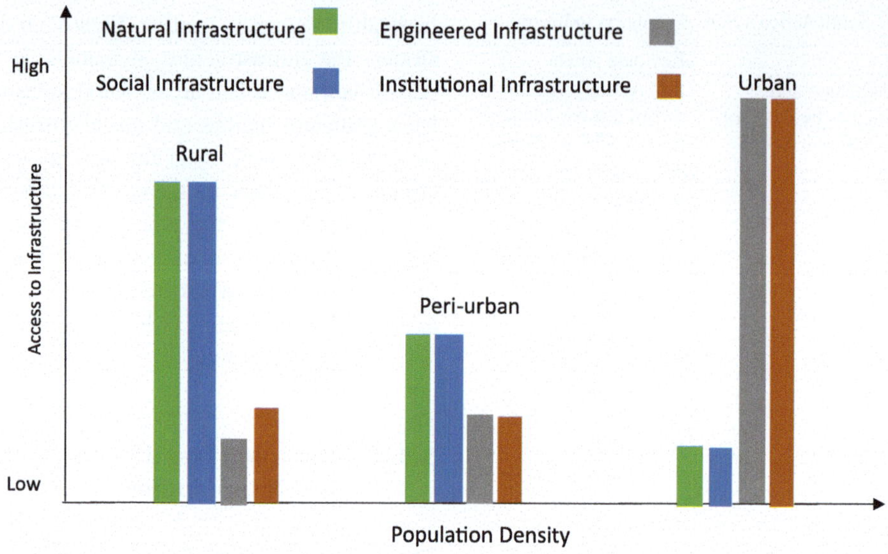

Fig. 5.6 Infrastructure dynamics in rural, peri-urban, and urban areas. (Source: Adapted from Hutchings et al. 2022)

menting tools such as HiAP and health impact assessments (HIA) can help achieve SDGs related to urban public health. These tools apply globally to all cities, including less developed countries (Ramirez-Rubio et al. 2019). An Australian case study by Schram et al. (2018), discussed in the previous section, may apply to countries implementing public digital infrastructure to ensure that provision is rolled out equitably. South Africa is moving toward digital infrastructure and may employ this strategy to improve urban public health.

5.3.7 Lessons Learned from COVID-19

COVID-19 catalysed technological innovation in urban public health (Umair et al. 2021). It prompted the implementation of IoT in the health sector and across sectors such as town planning, transportation, industry, and smart housing/architecture (Umair et al. 2021). COVID-19 demonstrated that a lack of advanced urban public health infrastructure may be the difference between the quick containment of an outbreak and a global pandemic. Most of the contact tracing systems used during the COVID-19 pandemic were devised and implemented as the pandemic progressed, and even though they proved valuable, they were not quick enough to prevent mass loss of life. Furthermore, early detection systems can significantly strengthen our capacity to contain future outbreaks (Yousif et al. 2021). It is thus worth investing in advanced infrastructure and technologies and improving existing ones to stay ahead of threats to urban public health.

5.4 A Localised Perspective of Infrastructure and Innovation

Bloemfontein, the city of roses, is no longer a visual representation of this slogan as it is currently characterised by overflowing pollution, sewerage, dilapidated roads and subpar urban public health infrastructure. The city is infrastructurally failing, as inferred by local news outlets such as OFM, Bloemfontein Courant, and the Daily Maverick, which continually headline infrastructure failures in the city, including roads, water, electricity, refuse and sewage. This section reviews the infrastructure status established by the Mangaung Metropolitan Municipality

(MMM), then reflects on these challenges outlined by primary data collected from Bochabela, Turflaagte, Phase 2 and Phomolong. These four settlements on the southeast end of the city house middle to low-income residents. The results from key informant interviews from the current realities of the marginalised communities living in these townships are analysed.

5.4.1 Mangaung Metro Municipality Service Outlook

The provision of basic infrastructure services is critical for preserving urban public health, as alluded to throughout this chapter. The municipality recognised that the area is characterised by high infrastructure backlogs and ageing infrastructure (MMM Metro One Plan 2022). This section assesses the MMM's infrastructure commitments and contrasts these with the services in the metro. Attention is afforded to electricity, water, sanitation, waste removal and roads, which the MMM Updated Status Quo Assessment from August 2023 acknowledges is subpar due to financial constraints.

5.4.1.1 Water

Water provision is a priority for the municipality, and the MMM Metro One Plan (2022) established that continued provision requires coordination, management and investment. While water is not the most critical backlog in the metro, extending water services still needs attention, as 17,555 households remain without water. According to the Service Delivery and Budget Implementation Plan (SDBIP) 2022/2023, the municipality plans 1884 new service connections. However, the shortfalls from previous municipal reports prove this goal may not be attained. Maintaining new and old service connections is critical for ensuring the quantity and quality of water provision. Unfortunately, this takes a long time, with only 20% of water call-outs being attended to within the municipal required 24-hour period. The lengthy call-out time results in large quantities of water waste that can be aligned with the increasing unplanned service interruptions, which the municipal SDBIP 2022/3 (2023) reports as occurring twice a month.

The quality and safety of water provided by the Bloemfontein water service system cannot be guaranteed, according to the Blue Drop Progress Report (2022), which found incongruence in the municipal sample reports. This aligns with the findings of the 2023/4 MMM SDBIP, which noted that the percentage of drinking water samples complying with SANS241 could not be confirmed as data is incongruent, a significant concern for urban public health. Additionally, the MMM Updated Status Quo Assessment (2023) alludes that most of the water in the metro is not fit for consumption, with two critical and high-risk supply systems. At best, Bloemfontein, Dewetsdorp and Thaba Nchu have medium-risk blue drop ratings. While part of the best in the metro, Bloemfontein forms part of 18% of the National Blue Drop Risk Rating (BDRR) profile in this category, as 48% of water supply systems nationally fall under the low-risk category (Blue Drop Progress Report 2022).

5.4.1.2 Sanitation

In 2021, the municipal green-drop score was 33%, an alarming 46% lower than the score just ten years ago, which was 79% (MMM Updated Status Quo Assessment 2023). Managing water purification and reticulation is essential as it aligns with water quality. A reported 60% of call-outs are attended to within 24 hours for sanitation/wastewater, meaning 40% of the time, wastewater and sewerage remain overflowing in Mangaung. However, this only applies to the 80% of Bloemfontein households with access to essential sanitation services.

While 92% of the municipal sanitation assets are in good or very good condition, leaks are persistent, and the maintenance budget is insufficient (MMM Updated Status Quo Assessment 2023). Around 79% of the households in MMM have access to flush toilets; this includes communal toilets shared by numerous families in the most marginalised communities. The percentile representation tends to diminish the extent of the 20% of households left below the service lines, as more than 50,000 ventilated improved pit (VIP) toilets are still found in Mangaung, all serving low-income areas and marginalised communities. Pit toilets pose a significant risk to urban public health, with 1400 households in the metro forced to use them. In 2020, the MMM Annual Performance Report reflected a Budget of 37 million to eradicate the bucket system in Bloemfontein; however, no clear commitment is reflected, and performance remains unclear.

5.4.1.3 Roads

Transportation networks within the city are almost entirely dependent on roads, as this is Mangaung's leading mode of transportation. Other services, such as stormwater and wastewater, are also linked to the efficiency of road infrastructure. The construction, maintenance and upgrading of all municipal roads are thus essential. However, road infrastructure is poorly maintained, resulting in poor connectivity and accessibility in some areas of the metro infrastructure (MMM Metro One Plan 2022).

Roads are arguably the worst infrastructure in the metro, as there is a continuous backlog in maintenance, potholes and resurfacing of existing roads. This is accompanied by failing traffic control measures, inconsistent road signs, and markings directly affecting access. The MMM Updated Status Quo Assessment (2023) further notes that the municipality is plagued with blocked drains and flooding stormwater, which creates a conducive environment for bacteria and algae, causing risk to urban public health. Considering the potential health risk, it is concerning that the municipal backlog for roads is the highest among all other infrastructure, at a staggering 56.77%, while 62.86% of the municipal road assets are in poor condition (MMM Updated Status Quo Assessment 2023).

As depicted in Fig. 5.7, the amount of road infrastructure repairs and maintenance in the metro is insufficient, with a 43% shortfall between what is required and the current reality; it is not surprising that the municipal roads are dilapidated. Importantly, funding is not the only hindrance to road maintenance in Mangaung, as challenges point directly to mismanagement, possibly even negligence. The municipality reports a 60% success rate of repairs done on reported pothole complaints within standard municipal response time (SDBIP 2023/2024). The initial concern with these stats is that 40% of the municipal roads remain unattended for extended periods. Another obstacle is that the

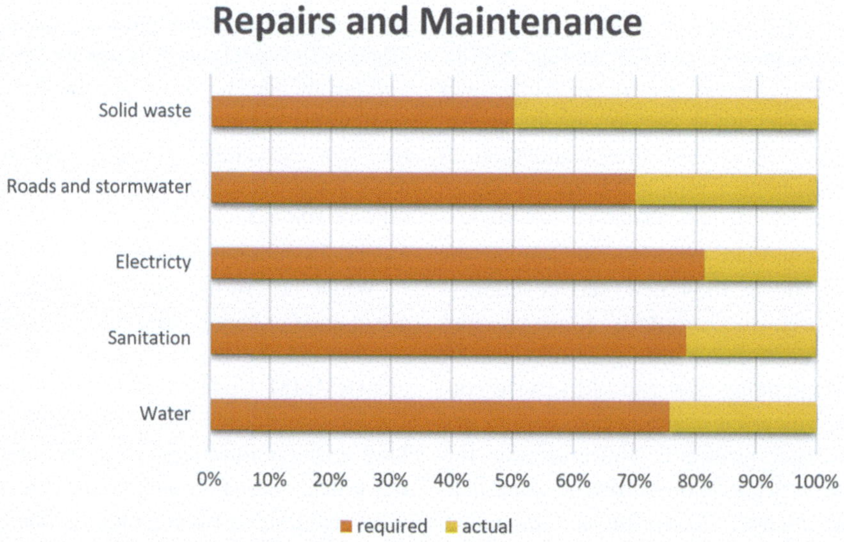

Fig. 5.7 Required and actual infrastructure repairs and maintenance in Mangaung Metropolitan Municipality. (Source: Adapted from MMM Updated Status Quo Assessment 2023)

document does not specify the 'standard municipal response time' limit, jeopardising accountability.

5.4.1.4 Energy and Electricity

The municipality has not substantially invested in renewable energy infrastructure; however, the traditional grid electricity infrastructure is fairly maintained. Electricity distribution is also widespread in most regions of the municipality, with a backlog of less than 5%. One of the most significant municipal electricity infrastructure challenges related to urban public health is street lighting, which directly affects safety, access and walkability in marginalised communities. While 34,733 public lights may seem sufficient, more than half of these lights serve middle to high-income suburbs. Street lighting and electrical infrastructure are often not fully maintained to ensure uninterrupted provision, leaving lower-income suburbs in the dark.

5.4.1.5 Maintenance

According to the MMM Updated Status Quo Assessment data from August 2023, 100% of the households on or above the minimum service levels have water and electricity. The same document, however, admits that this is not an accurate representation. Inaccurate data directly threatens the efficiency of infrastructure maintenance and repairs, as repairs cannot be ordered on faults that are not reported. Maintenance is not only an issue for electrical infrastructure in the metro, as defects remain undetected, unreported and unattended, especially in the townships.

The MMM Updated Status Quo Assessment (2023) posits poor project management has played a role in failing to achieve the IDP targets. Failure to meet IDP targets reduces grant funding, decreasing available funds for infrastructure development and leading to a lack of service delivery. The cyclical effects of mismanaged funds and project management failures go beyond the built environment and directly affect municipal infrastructure, particularly in marginalised communities with no access to privatised systems and services.

5.4.2 Existing Infrastructure and Challenges

A key aspect of resilient urban infrastructure is providing adequate and reliable access to essential services such as clean water, sanitation, refuse removal, healthcare facilities, roads, and public transportation. The quality and consistency of these services aid in maintaining good urban public health, with each type of infrastructure or service affecting residents' well-being.

While the physical health implications of inferior infrastructure are visible throughout the townships, the mental repercussions often go unseen. The psychological effects of being surrounded by degradation, dilapidation and decay cannot be denied. The helplessness and discouragement expressed by many residents in the interviews reveal that the administrative negligence of infrastructure in these marginalised communities goes beyond the surface. One of the respondents in the Bochabela location noted that the infrastructure was not well maintained, but good enough, considering that it is in the township, as she said:

> Here in the township, we cannot expect to have the same services as they have in town.

This sentiment was echoed by other residents, who considered themselves in acceptable circumstances compared to their families in more marginalised informal settlements.

This subsection examines the results of primary data collected in Bochabela, Turflaagte, Phase 2 and Phomolong as part of a collective study, which links several chapters of this book. The data from direct observations, key informant interviews and in-depth interviews were analysed to produce a transparent overview of the infrastructure in these marginalised communities. The research questions that directed the interviews for this chapter were used to detail the current state of infrastructure in these areas and the extent of perceived resilience.

1. What are the current urban public health challenges in the selected marginalised communities?

2. What is the state of the existing urban public health infrastructure and innovation in the marginalised communities?
3. What are the existing institutions responsible for urban public health infrastructure and innovation in the marginalised communities?
4. What are the new forms of infrastructure and innovation that can be implemented to improve the state of urban public health in these marginalised communities?

5.4.2.1 Water and Sanitation

The 2022 Bloem Water Annual Report recorded 40–48% water losses in the city, and the ageing infrastructure was a significant role player in these losses (Bloem Water 2021). Losses resulting from deteriorated infrastructure mean the end users receive limited water, and the municipality experiences financial loss. This phenomenon was prevalent in the settlements, where there were numerous puddles of water from leaks and infrastructure faults, as illustrated in Fig. 5.8. Another identifiable water challenge in the study areas was water puddles surrounding communal taps (see Fig. 5.8). In this case, the water loss results from taps left open due to resistant negligence and taps that do not close properly.

Municipal water leaks and sewerage have become a norm in many townships across Mangaung. There are streams of fresh water and puddles from burst pipes, which never run dry. Leaks are not only a waste of water and fiscal resources but also conducive to illness, causing bacterial growth and spreading waterborne diseases. Garbage collects in water puddles, which can rot and mould, as depicted in Fig. 5.9, thus increasing the risk of disease.

Sanitation infrastructure in the townships is varied, as the residents either had individual toilet facilities in-house, in their yards, or communal toilets shared by separate families. In Phomolong, many pit toilets remain in use as no municipal sanitation systems have been installed, as depicted in Fig. 5.10. According to Orner et al. (2018), pit toilets are either emptied or closed when full, and the health risks associated with both are significant. Further health risks are associated with using these toilets when there is insufficient ventilation or when they are old and used for extended periods (Odeku 2022).

In the settlement where formal toilets are constructed, overflow and leaks are visible as faecal sludge in septic tanks is not consistently extracted, and sewerage pipes burst without sufficient main-

Fig. 5.8 Water puddles around a communal tap and a large stagnant water puddle. (Source: Author)

Fig. 5.9 Garbage accumulating around a leaking water meter. (Source: Author)

Fig. 5.10 Sanitation structures. (Source: Author)

Fig. 5.11 Sewerage leaks. (Source: Author)

tenance. Some residents revealed that sewerage can overflow for weeks without municipal intervention, and they are forced to live with the stench. As depicted in Fig. 5.11, malfunctioning sanitation infrastructure results in toilet paper and faecal matter bursting and flowing in the yards and streets. The infrastructural failures expose communities to the fumes and bacteria associated with the sewerage.

5.4.2.2 Refuse Removal

The efficacy and consistency of garbage collection in Bloemfontein is a collective challenge. From the affluent north to the marginalised south, municipal refuse removal is inconsistent, insufficient and unacceptable. The residents in the four settlements identified refuse removal services as one of the significant service delivery challenges. Municipal plans for the settlements reflect refuse removal services accompanied by schedules. However, these services are scarce and inconsistent.

To a large degree, the community response to inconsistencies in refuse removal has been to dump in open spaces or road reserves. Garbage, building rubble, old food and even human waste (such as diapers and sanitary pads) are all piled up and left to decay, as depicted in Fig. 5.12. Children and livestock often wander into these garbage dumps, which creates exposure to bacteria and illnesses.

5.4.2.3 Roads

Developing walkable cities is essential in improving urban public health as it promotes a healthier lifestyle while simultaneously reducing carbon emissions. This means urban infrastructure should promote walking by developing sidewalks and ensuring sufficient street lighting. Low-income neighbourhoods often have minimal sidewalk lighting, encouraging crime, which decreases the walkability of these neighbourhoods by endangering the communities. As depicted in Fig. 5.13, the streets in Bochabela are not pedestrian-friendly as they do not have consistent sidewalks. The poor quality of the road and frequent potholes make it even more dangerous for pedestrians, as cars drive on the sidewalks to avoid potholes.

The study found that the settlements are accessible and have road infrastructure using municipal tar and gravel roads or informal roadways

Fig. 5.12 Garbage dumping. (Source: Author)

Fig. 5.13 Inconsistent sidewalks, Bochabela. (Source: Author)

extending the municipal roads. Most of the streets in the study areas presented low levels of resilience as they were not well built, and only the main roads were tarred, while the rest of the settlements had gravel or informal roadways. The gravel streets that meander through these townships are unusable during rainy seasons as they become muddy, and the run-off/absorption rate is slow (see Fig. 5.14).

Many roads are affected by frequent water leaks, which cause muddy puddles in the street, making driving difficult. This is an inconvenience, but it becomes an urban public health risk when accessibility for emergency vehicles such as ambulances and fire trucks is limited. With inconsistent road quality, minimal street names and barely any lights, these townships are not easily accessible in emergencies, even with the existing roads.

Travel time and distance to health facilities are significant factors for urban public health. Considering this, the communities were engaged in their experiences in this regard. Of the interviewed respondents, 60% presented discontent-

Fig. 5.14 Muddy roads Phomolong; Inconsistent road types, Bochabela. (Source: Author)

ment with the travelling time to clinics and the fact that they did not have these facilities in their townships. Residents must use public transport or walk more than 30 minutes to get to the clinics in neighbouring locations. This is particularly challenging for older people and those who are very ill. In Bochabela, the respondents noted that Pelonomi Hospital was close to them. However, it is constantly overcrowded as it is a regional hospital, so they, too, go to neighbouring townships for healthcare assistance in clinics.

5.4.2.4 Healthcare Facilities

Bloemfontein, the largest urban area in the Mangaung metro, has three public hospitals: Universitas Academic Hospital, National District Hospital and Pelonomi Regional Hospital. These hospitals serve not only the 598,383 people in the city but also a large proportion of the 861,651 people in the metro. When considering the physical land area of Bloemfontein, many individuals who use these hospitals have to travel notable distances to reach them. This is also true for the residents in the four neighbourhoods assessed in this study. Accessibility is, however, not the most significant challenge with urban public health facilities in the city. Many facilities are understaffed and structurally outdated while operating on minimal funds. According to the South African Institution of Civil Engineering (SAICE 2022), the Free State has the most significant number of high-risk health infrastructure conditions nationally. Malakoane et al. (2020) note that most Free State clinics are old, and their infrastructure is insufficient for the population size they need to serve.

There are no clinics in any of the four settlements in this study, meaning the residents depend on clinics in neighbouring townships. In Turflaagte and Phomolong, however, mobile clinics are assigned to provide residents with healthcare services. On paper, it is reported that they come on Wednesdays; however, the mobile clinics cannot be relied on as they are inconsistent. Sometimes, they arrive on other days of the week without prior communication with the residents, and sometimes, the mobile clinics do not come at all for extended periods.

5.4.3 Community Innovation in Local Infrastructure

The realities expressed in this section amplify the fact that the urban poor are being left behind in terms of service delivery, maintenance, and infrastructure. Municipal structures are not meeting the infrastructural needs of these marginalised communities, and not much innovation is being applied to them. As identified in the subsection on 5.4.2.2 Refuse removal, when the affluent experience service delivery and infrastructural failures, they privatise the systems. What are low-income settlements doing without the financial means to bridge the gap? This section visually explores the innovative and remedial actions taken and proposed by the residents in Phomolong, Bochabela, Phase 2 and Turflaagte as they relate to building resilience and infrastructure.

One of the community service interventions found in the townships was garbage collection systems. Community members with vans collect the rubbish from their neighbours at a fee per bag, and they dispose of the waste in the municipal landfills. This is a necessary intervention as pollution, particularly from dumping, is prevalent in these neighbourhoods. The municipal shortfall has presented a business opportunity for one of the residents (Fig. 5.15), who has a truck and garbage collectors, functioning like formal refuse removal services for those who pay the fee. Recycling, practised by many urban poor in South Africa, is prevalent in all four study areas. While not all the recycled materials are collected in those townships, the one-ton woven polypropylene bags are typical, as depicted in Fig. 5.15.

Roads throughout the townships are subpar and not well maintained. This lack of maintenance directly threatens resilience in areas that are difficult to access. Figure 5.16 illustrates the small ways in which community members intentionally bring solutions to the problems they experience. While this is arguably not an innovative solution, it is precisely the kind of community intervention that starts to instigate change on a local level, bridging the gaps left by poor service delivery and infrastructure management. Some community members have learned to live with the environment as it is. They remain dependent on the government, as expressed in the interviews by residents waiting for municipal infrastructure interventions, while others have vowed to make a change.

The innovative actions led by community members should not replace municipal infrastructure development and maintenance. Specialists must engage communities through education and partnership to encourage combined efforts toward infrastructure resilience. Innovation must be accompanied by community education on the usefulness of new technologies.

> Technology will only make things worse because people won't know how to use that technology, and

Fig. 5.15 Community innovation and entrepreneurship. Recycling bags. (Source: Author)

Fig. 5.16 Community-filled potholes, Bochabela. (Source: Author)

> no one cares about technology when they are sick; they just need to get help and go home to rest.

The response given by a community member provides a critical perspective. Innovation should increase efficiency and relieve the burden of those needing assistance. If this is not happening, it may be misdirected resources.

5.5 Discussion and Policy Implications

The findings presented in this chapter reveal the pressing challenges faced by marginalised communities in Bloemfontein, South Africa, regarding urban infrastructure and its impact on urban public health. The infrastructure challenges identified in Bloemfontein are multi-faceted and have far-reaching consequences for urban public health. The inadequate provision of clean water, sanitation, refuse removal, healthcare facilities, and transportation infrastructure exacerbates existing health disparities. While the physical health implications are evident, the mental health effects of living in deteriorated environments cannot be overlooked. Residents' feelings of helplessness and discouragement reflect the more profound psychological impact of administrative negligence on marginalised communities. One striking aspect is residents' acceptance of subpar services in marginalised communities. Statements such as, "Here in the township, we cannot expect to have the same services as they have in town," reveal a resignation to inequity. This resignation is echoed when residents compare their situations to those in more marginalised informal settlements. The infrastructure quality and access disparities highlight the urgent need for a more equitable approach to urban development.

The high water losses in Bloemfontein, as indicated in the 2022 Bloem Water annual report, directly result from ageing infrastructure. These losses limit the residents' access to water and impose a financial burden on the municipality. The prevalence of water leakages, as observed in all the settlements, poses significant urban public health risks. Stagnant puddles and water sources caused by leaks become breeding grounds for disease-causing bacteria and contribute to the spread of illness. Additionally, the sanitation infrastructure in some areas, like Phomolong, remains inadequate, with pit toilets leading to further health risks. Inconsistencies in municipal refuse removal services are a shared challenge among the settlements. Residents' response to these inconsistencies often involves open, unsanitary dumping. Children are exposed to these hazards, revealing the urgent need for improved refuse removal services to protect urban public health.

Walkable cities are essential for promoting healthier lifestyles and reducing carbon emissions (Guan et al. 2020). However, low-income neighbourhoods in Bloemfontein often lack pedestrian-friendly infrastructure, as highlighted by municipal documents. Walkability is restricted, and there are safety risks due to poor road quality and frequent potholes. Infrastructure challenges, such as inconsistent road quality and minimal street lighting, affect emergency response, limiting accessibility during critical situations. Access to healthcare facilities is crucial in urban public health (Sivakumar et al. 2020). In Bloemfontein, the challenges go beyond accessibility, as many healthcare facilities are understaffed, structurally outdated, and operate with limited resources. Furthermore, the absence of clinics in the settlements necessitates residents to rely on neighbouring townships for healthcare services, often facing long travel times. The structural conditions of healthcare facilities in the Free State, as reported by the South African Institution of Civil Engineering, further exacerbate these challenges.

While the challenges presented are substantial, there is hope in the form of community-led innovation. Residents in Phomolong, Bochabela, Phase 2, and Turflaagte have taken it upon themselves to address some of these issues. Community-led garbage collection systems and recycling initiatives have emerged as effective interventions. These initiatives address environmental concerns and create opportunities for employment and income generation, demonstrating the resourcefulness and resilience of these communities. However, while valuable, community-led innovation should not be a substitute for comprehensive municipal infrastructure development and maintenance. Education and partnerships between specialists and communities are vital to ensure effective and sustainable innovations. As one community member rightly pointed out, technology alone may not be the answer when immediate health needs must be addressed.

Overall, the findings of this study underline the urgent need for comprehensive infrastructure improvements in Bloemfontein, particularly in marginalised communities. Policy recommendations to bridge the existing gaps include equitable resource distribution, investment in infrastructure development, and addressing administrative negligence. Collaboration between communities, specialists, and municipal authorities is essential to build resilient urban infrastructure that supports urban public health. To adequately address the persistent infrastructure gap and improve urban public health, a multi-pronged approach is needed. Investment in resilient infrastructure must be prioritised, considering different communities' specific needs and vulnerabilities instead of a generic blueprint process. This includes developing and maintaining adequate water and sanitation systems, reliable waste management services, accessible healthcare facilities, and well-structured road networks.

Resilient infrastructure can withstand shocks, ensure uninterrupted services, and minimise health risks even in challenging circumstances. Moreover, the concept of smart cities should be harnessed to integrate technology-driven solutions that can enhance urban public health. Early detection systems and data-driven planning can facilitate efficient resource allocation, timely responses to health threats, and improved service delivery. However, technological innovation must be accompanied by community education and capacity building to ensure its effective utilisation, addressing concerns raised by residents.

Effective urban planning and government policy frameworks are indispensable for promoting resilient urban infrastructure and improving urban public health outcomes. Governments and local planning authorities must prioritise urban public health in urban planning and ensure that infrastructure development is equitable and inclusive. Urban planning and development policies should enforce accountability for maintenance and regular upgrades, preventing the decay and neglect witnessed in marginalised communities. Adopting a Health in All Policies approach can be instrumental in this endeavour. This approach requires a comprehensive assessment of the health implications of various policies and decisions, ensuring that urban development aligns with health objectives. Moreover, integrating health impact assessments into infrastructure projects can proactively identify potential health risks and inform design choices.

Financing resilient infrastructure remains a significant challenge, especially in low-resource settings. The Bloemfontein case study highlights the urgent need for increased financial allocation to infrastructure development, requiring innovative solutions to the existing shortfalls in infrastructure maintenance. In collaboration with international agencies, governments should explore innovative funding mechanisms, public-private partnerships, and grants to bridge the financial gap. Additionally, mobilising local resources and community involvement in infrastructure initiatives can enhance sustainability.

Regulatory and legal barriers often hinder the implementation of resilient infrastructure projects. Streamlining regulatory processes, removing bureaucratic obstacles, and enforcing compliance with standards are critical steps. Regulatory frameworks should also incorporate flexibility to accommodate emerging technologies and adapt to changing health needs.

Finally, addressing the infrastructure challenges highlighted in this chapter is not just a matter of urban development but urban public health and social equity. The resilience of marginalised communities in the face of adversity is commendable, but it should not be the sole solution. It is the collective responsibility of all stakeholders to work toward urban settlements where all residents, regardless of their socioeconomic status, enjoy the benefits of reliable infrastructure and improved urban public health.

5.6 Conclusion

In examining the intersection of urban infrastructure and urban public health challenges in Bloemfontein, South Africa, this study illuminated the relationship between the quality of urban infrastructure and the well-being of marginalised urban communities. The findings of this chapter revealed the urgency of addressing infrastructure disparities that persist in urban communities. Through our case study, Bloemfontein emerged as a microcosm reflecting the broader global issue of how inadequate infrastructure perpetuates health disparities, shaping both physical and psychological health outcomes for its residents.

The evidence presented in this study emphasises the pressing need to rectify infrastructure disparities in marginalised urban areas. These disparities are strongly linked to heightened health risks for urban residents, impacting the very essence of their lives. Beyond the physical ramifications, inadequate infrastructure has profound psychological implications, casting a shadow over the overall well-being of our communities. This study serves to encourage action from all stakeholders. Governments, local authorities, international agencies, and communities must unite to address infrastructure disparities to improve the overall urban public health and well-being of marginalised urban communities. All relevant stakeholders must commit to long-term investments in resilient urban infrastructure to protect citizens and empower them to thrive. The lessons from Bloemfontein resonate beyond its borders, reminding us that a healthier and more equitable urban future begins with dedication to resilient infrastructure and innovation.

As we continue to improve urban public health through infrastructure development, we recognise the need for ongoing research and evaluation. Future research should focus on assessing the effectiveness of policy interventions, measuring the impact of technology-driven solutions, and monitoring the long-term health outcomes brought about by infrastructure improvements. Advancing our understanding and commitment to this critical intersection of urban infrastructure and urban public health contributes to an improved and healthier urban future.

References

Abrams AL, Carden K, Teta C, Wågsæther K (2021) Water, sanitation, and hygiene vulnerability among rural areas and small towns in South Africa: exploring the role of climate change, marginalisation, and inequality. Water 13(20):2810

Abuya DO (2020) Management of the effects of land use changes on urban infrastructure capacity: a case study of Ruaka Town, Kiambu County, Kenya. Doctoral dissertation, University of Nairobi

Adewole A, Struthers JJ (eds) (2018) Logistics and global value chains in Africa: the impact on trade and development. Springer

Allam Z, Jones DS (2020) February. On the coronavirus (COVID-19) outbreak and the smart city network: universal data sharing standards coupled with artificial intelligence (AI) to benefit urban health monitoring and management. In: Healthcare (Vol. 8, No. 1). MDPI, p 46

Azzopardi-Muscat N, Brambilla A, Caracci F, Capolongo S (2020) Synergies in design and health. The role of architects and urban health planners in tackling key contemporary public health challenges. Acta Bio Medica: Atenei Parmensis 91(Suppl 3):9

Bhanye J, Dzingirai V (2020) Plural strategies of accessing land among peri-urban squatters. Afr Black Diaspora: Int J 13(1):98–113

Bhanye J, Dzingirai V (2022) Structures and networks of accessing and securing land among peri-urban squatters: the case of Malawian migrants at Lydiate informal settlement in Zimbabwe. African Identities 20(3):237–256

Biglieri S, De Vidovich L, Keil R (2021) City as the core of contagion? Repositioning COVID-19 at the social and spatial periphery of urban society. Cities Health 5(sup1):S63–S65

Bloem Water (2021) Bloem water annual report 2021/22. Bloem Water, Bloemfontein

Blue Drop Progress Report (2022) South Africa. Department of Water and Sanitation. 2022. Pretoria: Government printer.

Capolongo S, Rebecchi A, Buffoli M, Appolloni L, Signorelli C, Fara GM, D'Alessandro D (2020) COVID-19 and cities: from urban health strategies to the pandemic challenge. A decalogue of public health opportunities. Acta Bio Medica: Atenei Parmensis 91(2):13

Carbonell L, Hofmann P, Srikissoon N, Campos LC, Mbatha S, Lakhanpaul M, Mabeer V, Steenmans I, Parikh P (2023) Localisation of links between sanitation and the sustainable development goals to inform municipal policy in eThekwini municipality, South Africa. World Dev Sustain 2:100038

Chester MV, Allenby B (2019) Toward adaptive infrastructure: flexibility and agility in a non-stationarity age. Sustain Resilient Infrastruct 4(4):173–191

Cohen B (2015) Urbanisation, City growth, and the new United Nations development agenda. Cornerstone 3(2):4–7

Corvalan C, Villalobos Prats E, Sena A, Campbell-Lendrum D, Karliner J, Risso A, Wilburn S, Slotterback S, Rathi M, Stringer R, Berry P (2020) Towards climate resilient and environmentally sustainable health care facilities. Int J Environ Res Public Health 17(23):8849

D'Alessandro D (2020) Urban public health, a multidisciplinary approach. Urban Health: Participatory Action-research Models Contrasting Socioeconomic Inequalities in the Urban Context 24, pp.1–8

Economic Commission for Latin America and the Caribbean (ECLAC) (2021) Digital technologies for a new future (LC/TS.2021/43), Santiago, 2021

Fofana I, Chitiga-Mabugu M, Mabugu R (2018) South Africa milestones to achieving the sustainable development goals on poverty and hunger, vol 1731. Int Food Policy Res Inst

Forcellini D (2023) The role of redundancy of infrastructures on the seismic resilience (SR) of sustainable communities. Sustain For 15(15):11849

Francis R, Bekera B (2014) A metric and frameworks for resilience analysis of engineered and infrastructure systems. Reliab Eng Sys Safety 121:90–103

Franco IB, Chatterji T, Derbyshire E, Tracey J (2019) Actioning the global goals for local impact. Springer, Berlin, Germany

Gaal HO, Afrah NA (2017) Lack of infrastructure: the impact on economic development as a case of Benadir region and Hir-shabelle, Somalia. Dev Count Stud 7(1)

Gatzweiler F, Fu B, Rozenblat C, Su HJJ, Luginaah I, Corburn J, Boufford JI, Vela Valdes J, Nguendo-Yongsi B, Howden-Chapman P, Singh RB (2021) COVID-19 reveals the systemic nature of urban health globally. Cities Health 5(sup1):S32–S36

Goal 9: Industry, innovation and infrastructure—SDG tracker (2020) Our world in data. Available at: https://sdg-tracker.org/infrastructure-industrialization (Accessed 12 June 2023)

Guan C, Keith M, Hong A (2020) Designing walkable cities and neighborhoods in the era of urban big data. Urban Plan Inter 34(5)

Haywood LK, Funke N, Audouin M, Musvoto C, Nahman A (2019) The sustainable development goals in South Africa: investigating the need for multi-stakeholder partnerships. Dev South Afr 36(5):555–569

Hugo M, Bhanye J (2022) The influence of classical' green' theories on contemporary industrial planning practices: a review study. Town Reg Plan 81:39–52

Husain A, Akram M (2022) Dimensions of Well-being. Psycho Inf Technol

Hutchings P, Wilcock S, Lynchn K, Bundhoo D, Brewer T, Cooper S, Keech D, Mekala S, Mishra PP, Parker A, Shackleton CM, Venkatesh K, Vicario DR, Welivita I (2022) Understanding rural-urban transitions in the Global South through peri-urban turbulence. Nature Sustainability 5:924–930

Kynčlová P, Upadhyaya S, Nice T (2020) Composite index as a measure on achieving Sustainable Development Goal 9 (SDG-9) industry-related targets: The SDG-9 index. Appl Energy 265:114755

Lennon M, Douglas O, Scott M (2017) Urban green space for health and Well-being: developing an 'affordances' framework for planning and design. J Urban Des 22(6):778–795

Luken RA, Saieed A, Magvasi M (2022) Industry-related sustainable development Goal-9 progress and performance indices and policies for sub-Saharan African countries. Environ Dev 42:100694

Malakoane B, Heunis JC, Chikobvu P, Kigozi NG, Kruger WH (2020) Public health system challenges in the Free State, South Africa: a situation appraisal to inform health system strengthening. BMC Health Serv Res 20:1–14

Mangaung Metropolitain Municipality (2023) Revised Service Delivery and Budget Implementation Plan (SDBIP) 2022/2023. Bloemfontein, Office of the City Manager IDP and OPM

Mangaung Metropolitain Municipality (2023) Revised Service Delivery and Budget Implementation Plan (SDBIP) 2023/2024. Bloemfontein, Office of the City Manager IDP and OPM

Mangaung Metropolitan Municipality (2023) Updated Status Quo Assessment. Bloemfontein, Aug.

Matamanda AR, Dunn M, Nel V (2022) Broken bridges over troubled waters: COVID-19 and the urban poor residing in Dinaweng informal settlement, Bloemfontein, South Africa. S Afr Geogr J 104(3):309–327

Meerow S, Pajouhesh P, Miller TR (2019) Social equity in urban resilience planning. Local Environ 24(9):793–808
MMM Metro One Plan (2022) Mangaung Metropolitain Municipality. 2022. Bloemfontein.
National Planning Commission (2012) National Development Plan Vision 2030. Available online at: http://www.gov.za/issues/national-development-plan-2030 (Accessed 15 June 2023)
NU CEPAL (2019) The 2030 agenda and the sustainable development goals: an opportunity for Latin America and the Caribbean. Goals, targets and global indicators
Odeku OK (2022) Critical analysis of school pit toilet system as an impediment to the right to access quality education in South Africa. Afr J Public Affairs 13(1):97–109
OECD (2017) Meeting of the OECD Council at Ministerial Level
OECD (2018) Bridging the rural digital divide. OECD Digital Economy Papers, No. 265, OECD Publishing, Paris, https://doi.org/10.1787/852bd3b9-en (Accessed 28 July 2023)
Olszewska-Guizzo A, Fogel A, Benjumea D, Tahsin N (2022) Sustainable solutions in urban health: transdisciplinary directions in urban planning for global public health. In: Sustainable policies and practices in energy, environment and Health Research: addressing cross-cutting issues. Springer, pp 223–243
Orner KD, Naughton C, Stenstrom TA, (2018) Pit toilets (latrines). Water and sanitation for the 21st century: health and microbiological aspects of excreta and wastewater management (Global Water Pathogen Project)
Palmer I, Graham N, Swilling M, Robinson B, Eales K, Fisher-Jeffes L, Käsner SA, Skeen J (2016) South Africa's urban infrastructure challenge. In: Department of cooperative governance and traditional affairs, pp 1–41
Pozoukidou G, Chatziyiannaki Z (2021) 15-Minute City: decomposing the new urban planning eutopia. Sustain For 13(2):928
Puplampu KP, Hanson KT, Shaw TM (2017) From MDGs to SDGs: African development challenges and prospects. In: From millennium development goals to sustainable development goals. Routledge, pp 1–10
Ramirez-Rubio O, Daher C, Fanjul G, Gascon M, Mueller N, Pajín L, Plasencia A, Rojas-Rueda D, Thondoo M, Nieuwenhuijsen MJ (2019) Urban health: an example of a "health in all policies" approach in the context of SDGs implementation. Glob Health 15:1–21
Republic of South Africa (1996) Constitution of the Republic of South Africa, act 108 of 1996. Government Printer, Pretoria
Schram A, Friel S, Freeman T, Fisher M, Baum F, Harris P (2018) Digital infrastructure as a determinant of health equity: an Australian case study of the implementation of the National Broadband Network. Aust J Public Adm 77(4):829–842
Sivakumar I, Manimekalai K, Ranjithkumar A (2020) Accessing public health facilities: rural and urban disparities. J Crit Rev 7(3):382–388
South African Institution of Civil Engineering (2022) SAICE 2022 infrastructure report card for South Africa. Hartwood Digital Printing, Midrand
Stats SA (2019) Sustainable development goals: country report 2019. Statistics South Africa
Sustainable Development Solutions Network (2022) Sustainable development report. From crisis to sustainable development: the SDGs as Roadmap to 2030 and beyond. https://dashboards.sdgindex.org/static/profiles/pdfs/SDR-2022-south-africa.pdf (Accessed 6 June 2023)
Tomaselli MF, Timko J, Kozak R, Bull J, Kearney S, Saddler J, Zhu X (2019) SDG 9: industry, innovation and infrastructure–anticipating the potential impacts on forests and forest-based livelihoods. In: Sustainable development goals: their impacts on forests and people. Cambridge University Press, United Kingdom, pp 279–314
Umair M, Cheema MA, Cheema O, Li H, Lu H (2021) Impact of COVID-19 on IoT adoption in healthcare, smart homes, smart buildings, smart cities, transportation and industrial IoT. Sensors 21(11):3838
United Nations (2020) The sustainable development goals report 2020. United Nations, New York, NY
United Nations (2022) The sustainable development goals report 2022. United Nations, Nairobi
United Nations Department of Economic and Social Affairs (2015) Transforming our world: the 2030 agenda for sustainable development. United Nations, New York, NY
United Nations Department of Economic and Social Affairs (2017) World population prospects: the 2017 revision. United Nations, New York, NY
United Nations in South Africa (2023) Our work on the sustainable development goals in South Africa. https://southafrica.un.org/en/sdgs (Accessed 30 July 2023)
United Nations Development Programme (2015) Human Development Report 2015. United Nations: New York, NY, USA
Van de Pas R, Hill PS, Hammonds R, Ooms G, Forman L, Waris A, Brolan CE, McKee M, Sridhar D (2017) Global health governance in the sustainable development goals: is it grounded in the right to health? Global Chall 1(1):47–60
Wernecke B, Mathee A, Kunene Z, Balakrishna Y, Kapwata T, Mogotsi M, Sweijd N, Minakawa N, Wright CY (2021) Tracking progress towards the sustainable development goals in four rural villages in Limpopo, South Africa. Ann Global Health 87(1)
World Health Organization (1989) Constitution. Am J Public Health 36(11):1315–1323
Wu J, Xie X, Yang L, Xu X, Cai Y, Wang T, Xie X (2021) Mobile health technology combats COVID-19 in China. J Infect 82(1):159–198
Yannis G, Chaziris A (2022) Transport system and infrastructure. Transportation Research Procedia 60:6–11
Yousif M, Hewage C, Nawaf L (2021) IOT technologies during and beyond COVID-19: a comprehensive review. Future Internet 13(5):105
Zimmerman R (2005) Mass transit infrastructure and urban health. J Urban Health 82:21–32

Liveability Assessment in South Africa's Hostel Accommodation: Implications for Urban Health and Sustainable Development Goal 11

6

Lucia Leboto-Khetsi, Jennilee Kohima,
Tazviona Richman Gambe,
Thulisile N. Mphambukeli, and Siphokazi Rammile

6.1 Introduction

The Sustainable Development Goals (SDGs) provide a global framework to address social, economic, and environmental challenges, with SDG11 focusing on inclusive, safe, resilient, and sustainable cities and human settlements (United Nations 2015). Urban health, involving the well-being of urban dwellers, is a critical concern, especially in marginalized communities facing housing and service access issues (World Health Organization 2016). SDG 11 is connected to the concept of urban health, which is influenced by urban planning and liveability. The goal of SDG 11 is to foster sustainable urban planning, with a focus on providing affordable housing, efficient transportation, green spaces, and adequate infrastructure. Moreover, SDG 11's objective to reduce environmental pollution and improve waste management aligns with addressing poor sanitation and pollution, which impact public health. Furthermore, The emphasis of SDG 11 on building urban resilience is essential for mitigating health impacts during disasters, requiring improved warning systems, healthcare infrastructure, and preparedness (United Nations 2015). Incorporating SDG 11 principles into urban planning and designing urban areas with care can significantly improve the health and well-being of residents. This can be achieved by promoting physical activity and reducing exposure to hazards, resulting in liveable, healthier, and more sustainable environments for inhabitants (World Health Organization 2016).

Cities exhibit a diverse range of living conditions and human experiences (Galea and Vlahov 2005). The SDGs offer an opportunity to create just, equitable, economically prosperous, and resilient cities, which withstand environmental changes. Hostel accommodation, established during apartheid to segregate and control black South Africans, typically lacks essential services, leading to health challenges such as infectious diseases, inadequate nutrition, mental health

L. Leboto-Khetsi (✉) · T. R. Gambe
Centre for Development Support, University of the Free State, Bloemfontein, South Africa

J. Kohima
Department of Architecture and Spatial Planning, Namibia University of Science Technology, Windhoek, Namibia

T. N. Mphambukeli
Department of Urban and Regional Planning, University of Johannesburg, Johannesburg, South Africa

S. Rammile
Department of Urban & Regional Planning, University of the Free State, Bloemfontein, South Africa

A. R. Matamanda, V. Nel (eds.), *Sustainable Development Goals and Urban Health*, Sustainable Development Goals Series, https://doi.org/10.1007/978-3-031-68734-1_6

issues, and limited healthcare access (Lemon 1991; Strauss 2019). By examining the interplay between SDG 11 and urban health in this context, this study explores the complex dynamics between urban development, health outcomes, and social justice, while suggesting potential pathways for sustainability. This study contributes to understanding the link between sustainable cities and urban health while offering effective strategies to address inequalities and improve health outcomes for marginalized populations. The findings will guide policymakers, urban planners, and public health practitioners in implementing interventions that align with SDG 11 objectives, promoting sustainable urban development, equitable access to services, and fostering healthier and liveable cities.

This chapter explores the links between SDG 11 and urban health, focusing on a specific case study of Hostel 1, an apartheid-era dormitory in South Africa under Mangaung Metropolitan Municipality. The housing and service delivery conditions in Hostel 1 exemplify the challenges faced by marginalized populations, where apartheid planning influenced current poor living conditions and limited access to basic services (Newton and Schuermans 2013; Sekoboto and Landman 2018). The study aims to understand this link by examining urban health through the lens of social services and housing conditions in Hostel 1, with four objectives.

1. To explore and understand the lived experiences and perspectives of residents living in a South African apartheid hostel about SDG 11 and its relevance to addressing the housing and health issues in their community.
2. To investigate and document the specific challenges that residents face regarding housing, infrastructure, and access to services in the dormitory.
3. To examine the impact of these challenges on the health and well-being of residents in the context of SDG 11.
4. To generate insights and recommendations for strategies and interventions to improve housing conditions, infrastructure, and access to services for sustainable development and improved health outcomes.

The chapter outlines the objectives, targets, and indicators of SDG 11 concerning the multidimensional nature of liveability and considers how social determinants and living conditions impact health outcomes. The historical and socio-economic factors shaping the South African urban context are discussed, along with relevant policies and interventions. The study's findings are linked to existing literature, offering policy implications and recommendations for urban planning strategies to improve health outcomes and promote sustainable development. The chapter concludes with key findings, study limitations and suggestions for future research.

6.2 Conceptual Underpinnings

In line with SDG 11, which focuses on inclusive, safe, resilient, and sustainable cities and human settlements, the concept of liveability guides this study. The definition of liveability is still subject to debate as no universally acceptable definition exists (Sarkarb and Bardhana 2020; Sochacka et al. 2021). In simple terms, liveability concerns "the ability of living space to support well-being or quality of life…" (Sarkarb and Bardhana 2020: 1). The concept can be used to access living standards despite being conceptualised differently across cities in different world regions (Paul and Sen 2020). Liveability has diverse applications, including transport studies and urban water management (Sochacka et al. 2021). In African cities, governance plays a crucial role in determining liveability, while cultural heritage and city image are the least significant determinants (Lawanson et al. 2013). It is worth noting that experiences vary from one city to another.

Understanding liveability and its indicators or determinants depends on the discipline and context in which it is used. Liveability can be viewed through the socio-physical lens. Social liveability indicators include privacy, safety, security and social cohesion, while physical liveability indicators include air quality, ventilation, and thermal comfort. These indicators are affected by the built environment in different ways, which has a bearing on urban health (Sarkarb and Bardhana 2020). According to Skalickya and Čerpesb (2019), residential environment liveability can be assessed through key aspects, including environmental, sociological, functional and cultural (Table 6.1).

This study conceptualizes liveability at a micro level, focusing on workers' hostels constructed in pre-independence South Africa. We assess how their state of repair, level of service provision, governance and functionality influence the well-being of the residents. Specifically, how Hostel 1 residents' social, physical, psychological, economic and spiritual well-being is affected by the building's condition, level of service provision, governance and the hostel's functionality.

Urban health is a linked concept that is essential in this study. Simply defined, urban health encompasses the influence of the physical and social urban environment on the health and well-being of the urban population. The built environment and accessibility of health and social services in urban areas can impact physical and mental health. Empirical evidence highlights the prevalence of conditions such as asthma, respiratory issues, injuries, psychological distress, and child development issues because of these determinants (Galea and Vlahov 2005; Wuerzer 2014).

Table 6.1 Key aspects of residential environment liveability

Key aspect	Description
Environmental	Urban development processes aiming to promote residents' quality of life through environmental protection and ensuring urban health
Sociological	Urban environments should promote mental and physical well-being and socialisation
Functional	Residential environments should promote the coexistence of various social groups while preventing the isolation of some groups
Cultural	Residential environments should have an identity that attracts people and makes them want to identify with it

Source: Skalickya and Čerpesb (2019)

Therefore, various outcomes of urbanization, including acute shortages of municipal services, urban sprawl, congestion and the growth of slums, negatively affect urban health.

The built environment can influence urban health positively or negatively. On one hand, properly planned cities (as envisioned by SDG 11) promote urban health by providing recreational facilities and accessible services; factors that improve the well-being of communities. On the other hand, the supply of poor-quality water, lack of proper sewage management and air pollution can negatively affect urban health (Wuerzer 2014) as this impacts communities' social and economic well-being.

6.3 Measuring Progress Towards Sustainable Development Goal 11

Cities serve as hubs for activities such as commerce, culture, science, productivity, and social development (Szpak 2016). However, urbanization has brought challenges including health implications. To address these challenges effectively, efficient urban planning and management practices are essential. With over 50% of the global population currently residing in cities and an expected increase to nearly 70% by 2050, it is crucial to prioritize participatory platforms that allow inclusive achievement of SDG 11. The focus of SDG 11 is to enhance urban planning, management, and development, making urban spaces more inclusive, safe, resilient, and sustainable. This objective is significant for governments struggling to accommodate the rising urban population, especially in concentrated areas with extreme poverty (Chopde and Nichat 2013; Tanwar et al. 2018).

Urban areas offer opportunities, including access to basic services, infrastructure, jobs, land, and affordable housing (Opoko and Oluwatayo 2014; King et al. 2017). However, close to one billion urban poor live in slums and peri-urban informal settlements, with Southeast Asia and Africa having many such cases. This is problematic in African cities where the majority live in squalid conditions with related health and safety issues (Gambe et al. 2023). Besides being overcrowded, unplanned settlements are associated with poor waste management, contributing to landfill buildup and the generation of harmful greenhouse gases like methane, negatively impacting the environment and human health (Zimring and Rathje 2012). Hence, the scale and character of urbanization have a significant influence on urban health.

Since urban health issues in the North do not differ significantly from those in the South (Wuerzer 2014), cities in the latter can learn from the former. To assess progress towards SDG 11, specific targets and indicators (Table 6.2) have been established. These indicators help measure the effectiveness of projects and initiatives aimed at achieving SDG 11.

The United Nations SDGs are a set of 17 global targets adopted in 2015 to be achieved by 2030 (UN 2015). These goals are tracked using 232 unique indicators. Cities in different countries in the North and South have recorded different experiences. For example, in Spain, the City of Valencia has used ISO 37120 standards to develop key performance indicators (KPIs), enabling integrated and transparent decision-making across the administration (Hajduk 2020; Takiya et al. 2022). This supports indicators under SDG 11, particularly the goal of enhancing inclusive and sustainable urbanization. The standardized KPIs have facilitated the use of the Internet of Things (IoT) for data collection, addressing specific issues such as air quality and noise control (Ban 2016). By reporting ISO 37120 KPIs, Valencia can compare its performance with other compliant cities, use rational dashboards for decision-making, and align vertical IoT solutions to achieve quicker and standardized results. This is an example of how improving certain urban environment characteristics can promote urban health. The strategies in Valencia reduced air pollution, improved noise control, and enhanced transparency in municipal decision-making. These are positive factors that improve the well-being of urban communities.

Table 6.2 Sustainable Development Goal 11 targets and indicators

11.1 Safe and affordable housing
Ensure access for all to adequate, safe, and affordable housing and basic services and upgrade slums. Indicator: Proportion of urban population living in slums, informal settlements, inadequate housing.
11.2 Affordable and sustainable transport systems
Provide access to safe, affordable, accessible, and sustainable transport systems for all, with a focus on expanding public transport by 2030. Indicator: Proportion of the population with convenient access to public transport, by sex, age, and persons with disabilities.
11.3 Inclusive and sustainable urbanization
By 2030, enhance inclusive and sustainable urbanization and participatory human settlement planning and management. Indicators: Ratio of land consumption rate to population growth rate; proportion of cities with a direct participation structure of civil society in urban planning and management.
11.4 Protect the World's culture and natural heritage
Strengthen efforts to protect and safeguard cultural and natural heritage. Indicator: Total per capita expenditure on preservation, protection, and conservation of heritage.
11.5 Reduce the adverse effects of natural disasters
Significantly reduce the number of deaths, affected people, economic losses caused by disasters. Indicators: (1) number of deaths, missing persons, and directly affected persons attributed to disasters per 100,000 populations; (2) economic loss attributed to disasters concerning the GDP; (3) damage to critical infrastructure and disruptions to basic services attributed to disasters.
11.6 Reduce the environmental impact of cities
Reduce adverse per capita environmental impact of cities on air quality and waste management. Indicators: Proportion of municipal solid waste collected and managed in controlled facilities; annual mean levels of fine particulate matter (PM2.5 and PM10) in cities.
11.7 Provide access to safe and inclusive green and public spaces
By 2030, provide universal access to safe, inclusive, and accessible green and public spaces. Indicators: Average share of the built-up area of cities that is open space for public use, by sex, age, and persons with disabilities; Proportion of persons victim of physical or sexual harassment in the previous 12 months, by sex, age, disability status, and place of occurrence.

(continued)

Table 6.2 (continued)

11.A Strong national and regional development planning
Strengthen national and regional development planning to support positive links between urban, peri-urban, and rural areas by 2030. ***Indicator: Number of countries with national urban policies/ regional development plans responding to population dynamics, ensuring balanced territorial development, and increasing local fiscal space.***
11.B Implement policies for inclusion, resource efficiency, and disaster risk reduction:
By 2020, increase the number of cities adopting integrated policies and plans for inclusion, resource efficiency, mitigation and adaptation to climate change, and disaster risk reduction. ***Indicators: (1) number of countries with national disaster risk reduction strategies aligned with the Sendai framework; (2) proportion of local governments with local disaster risk reduction strategies aligned with national strategies.***
11.C Support least developed countries in sustainable and resilient buildings:
Support least developed countries financially and technically to build sustainable and resilient buildings using local materials.

Source: Our World in Data Team (2023)

In North America, the Common Alerting Protocol (CAP) standard (ITU X.1303) is used by the United States National Oceanic and Atmospheric Administration (N.O.A.A.) and other national administrations for public emergency alerts (Chronaki et al. 2011). The success of CAP is evident, with 75% of the global population residing in a nation with operational or under development CAP alert feeds. The CAP significantly contributes to achieving SDG 11.b by promoting holistic disaster risk management at all levels in line with the Sendai Framework for Disaster Risk Reduction 2015–2030 (Pecorella et al. 2015). Safety is, therefore, one of the critical aspects of both liveability and urban health achieved by the CAP. Urban communities in coastal regions need early systems to protect their well-being. These systems should be in place to limit the effects of floods on their livelihoods. Many urban communities have lost their properties or lives during floods due to a lack of early warning systems. Thus, early warning systems promote urban health.

A trial of the new International Standard ISO 22327 on "Security and resilience – Emergency management – Guidelines for implementation of a community-based landslide early warning system" was conducted in 98 districts across 28 provinces of Indonesia (Budiarti et al. 2020). This standard significantly improved community resilience to landslide disasters and directly contributed to achieving SDG 11.b on disaster risk management. Early warning systems were installed in high-risk districts, and technical assessments determined hazard risk areas (Gumiran and Daag 2021; Peters et al. 2022). Vital components of effective development (Fathani et al. 2016; ISO 22327 2018), and by extension urban health, were identified as risk assessment; socialization; forming disaster-prepared responder teams; operational evacuation guides; developing standard procedures; monitoring; early warning; evacuation rehearsals; and developing local authorities' and communities' commitment towards operation and management of landslide early warning systems. Educating communities on landslides and evacuation procedures improved urban health by reducing casualties caused by landslides (Gamperl et al. 2023; Fathani et al. 2023). For sustainable success, collaboration between the government and the community is crucial. The standard has become a vital guideline for disaster management agencies in Indonesia and was included in the National Medium-term Development Plan (2015–2019) to reduce landslide disaster risk. The case of Indonesia shows region-specific strategies that countries can adopt to promote urban health. Communities living in disaster-prone areas need to have their social, economic, and psychological well-being protected through measures implemented by central governments and local municipalities.

Egypt improved urban health by addressing traffic congestion and safety incidents through implementing ISO 14813-1:2015 standards on "Intelligent transport systems" for key highways and transportation routes. This supported SDG 11.2 to improve road safety and provide sustainable transport systems for all (Abdelaty et al. 2023; Elbeltagi et al. 2004; Hakim and Mohsen

2017). These standards facilitate strong interoperability between systems, avoiding vendor lock-in and ensuring improved services, quality and performance at cost-effective options. Furthermore, they promote future scalability by adopting a modular approach based on standardized architecture and enhanced maintenance planning. Egypt's efforts to improve systems operations within the built environment promote urban health by reducing road traffic accidents and congestion, which causes loss of business through delivery delays. An efficient transport system connects people and places, promotes hassle-free mobility, improves accessibility to services and reduces air pollution generated by exhaust fumes from cars stuck in congestion.

The American Society for Testing and Materials (ASTM) International Standards were adopted in Zambia to increase the recycling of building materials, particularly concrete, reducing landfill waste and conserving water in construction practices (Beck et al. 2018; Fisonga et al. 2019; Omary et al. 2009). These standards contribute to achieving SDG 11.1 and 1.6 by ensuring access to safe and affordable housing and reducing environmental impact in cities. Zambia's experience validates lower costs, environmental stewardship, and sustainable construction, promoting recycling in the construction industry. This approach reduces landfills and supports environmental preservation (Agboola et al. 2020; Huang et al. 2020). The implementation of these standards in multiple countries demonstrates their efficacy in promoting sustainable construction practices, environmental stewardship and urban health.

6.4 The Context of SDG 11 and Urban Health in South Africa

With over 67% of the South African population living in urban areas (O'Neill 2024), sustainable urban planning is crucial to address rapid urban growth rates. However, planning policies and legislation are struggling to keep up with the pace of urbanization, leading to poor service delivery, deteriorating infrastructure, and inadequate housing. Informal settlement development exacerbates the housing and health link in South Africa, as highlighted by Marais and Cloete (2014), who note a lack of in-depth research due to the prevailing focus on housing policy and informal settlements.

Urban informal settlements are susceptible to infectious diseases, but it is important to recognize that urbanization also contributes to non-communicable diseases like diabetes, obesity, and heart disease in sub-Saharan Africa (Matamanda et al. 2022). Urban health challenges vary across different urban environments, with informal and low-income settlements facing the most significant concerns. In 2020, the World Bank reported that over 24% of South African urban dwellers lived in informal settlements.

Unliveable urban conditions, including high density, poor housing, and inadequate infrastructure and services, directly impact urban health (Boyce et al. 2019). Negligence in urban planning leads to poor living conditions, affecting vulnerability to illness and recovery. This vulnerability is particularly pronounced in low-income, high-density, heat-prone areas with limited access to health infrastructure. Thompson, Matamale, and Kharidza (2012) highlighted that poor housing quality can increase respiratory diseases and malaria prevalence, while the availability of water, sanitation, and health services influences the distribution of diseases like cholera.

The SDG 11 indicators in South Africa show significant challenges, but there has been moderate improvement. The government has clustered the SDGs into social, environmental, economic, governance, peace, justice and security categories to enhance national attainability (Statistics South Africa 2019). Key targets prioritized from SDGs include only two from SDG 11, namely 11.1 (safe and affordable housing) and 11.3 (inclusive and sustainable urbanization) under environmental goals. Despite expectations of high success, the 2022 sustainable development report reflects a decline in performance for indicator 11.1.1, contrasting with the overall moderate improvement in SDG 11 (Sachs et al. 2022).

South Africa's country score on SDG indicators improved from 64% in 2019 to 72% in 2023. Despite the positive progress made towards achieving the SDGs in 2023, no data were available for six indicators, and only limited data were available for five indicators within SDG 11 in both 2019 and 2023 (Statistics South Africa 2019, 2023).

SDG 11.1.1, which concentrates on service provision for informal housing dwellers, aligns with the national Medium Term Strategic Framework (MTSF), but many targets for housing delivery and improved liveability set for 2019 have not been achieved. The 2022 SDG report highlights that 230 million people in sub-Saharan Africa still live in slums (UN 2022). On a municipal level, the Mangaung Strategic Development Framework (SDF) 2022/3 reflects that 11% of all dwelling units within the metro are informal; this is inclusive of informal settlements and backyard dwellings, with the Bloemfontein figures higher at 14%. The attainment of sustainable cities and communities requires national and local policy tracking. Key policy alignment for SDG 11 includes the National Development Plan (NDP) and the MTSF. Additionally, the Spatial Planning and Land Use Management Act (SPLUMA) plays a critical role in establishing sustainable, inclusive, safe, and resilient cities, despite not being specifically aligned with SDGs.

The 2019 Country Report shows that national policies listed in Table 6.3 are aligned with SDG targets and are essential for developing and maintaining human settlements. Despite the national focus on informal settlement upgrading, inclusive human settlements, and essential services, the delivery rate remains low.

The NDP emphasizes local integrated human settlement, prioritizing community participation and empowerment in planning and management. It addresses key aspects of SDG 11, such as bridging housing gaps, informal settlement upgrading, and ensuring accessible and affordable public transportation for all income groups. The Department of Planning, Monitoring, and Evaluation and the United Nations Development Programme (2019) highlight the intersections between the NDP and the SDGs (Table 6.4).

Table 6.3 National policies aligned with Sustainable Development Goals' targets

Legislation	Year	Regulates
Housing Act	1997	Support for adequate housing delivery
Water Services	1997	Water and sanitation provision
National Environmental Management: Waste Act	2008	Minimizing environmental impact of development and waste management services
Municipal Systems Act	2000	Governance and administration within municipalities and local basic service provision and management
Breaking New Ground	2004	Sustainable human settlement development and housing principles and norms
National Housing Code	2009	Housing development and informal settlement upgrading

Table 6.4 Connections between the National Development Plan (NDP) and Sustainable Development Goals (SDGs)

NDP area	SDG 11 target
Economic infrastructure (4)	11.2
Environmental sustainability and resilience (5)	11.5 & 11.6
Transforming human settlements (8)	11.1, 11.2, 11.3, 11.7 & 11.b

Locally, Municipal Integrated Development Plans (IDPs) drive SDG 11 targets, as they must include sector plans like Human Settlements, Disaster Management, Transport, and Water Services Plans, which align with the SDG. The 2022 SDG report reflected that municipalities in sub–Saharan Africa have the lowest solid waste collection rates, with an average collection of only 54.5%. This inefficiency is reflected nationally, as noted by Polasi, Matinise and Oelofse (2020); the success of municipal solid waste programmes is hindered by financial, managerial, operational, and legislative challenges. While solid waste collection is a mandatory service by the government, it is often inconsistent in most serviced areas and non-existent in the rest, especially in low-income settlements. Low-income communities included in municipal collection routes on paper often experience gaps in the collection, resulting in alternative waste disposal methods (Haywood et al. 2021). This results in two notable pollutants, fumes from burning waste or litter and illegal dumping. Contamination risks increase exponentially as garbage is strewn across the neighbourhoods, which exposes dwellers to illnesses. The National Environmental Management: Waste Act (No. 59 of 2008) is a critical policy guarding against the hazards of solid waste to prevent environmental degradation and urban health risks; however, the implementation rate of the policy remains subpar on a local level.

Inadequate access to safe drinking water, sanitation, and solid waste removal increases health risks in communities (Marais and Cloete 2014). South African informal settlements and low-income housing suffer from compromised ventilation and air quality due to congestion and pollution, leading to severe health implications and increased risk of infections, including tuberculosis (Weimann and Oni 2019). Air pollution remains a significant problem in the country, necessitating urgent intervention (Sachs et al. 2022). Proximity of low-income housing to healthcare facilities is crucial for effective interventions and mitigation of adverse conditions (Marais and Cloete 2014). Overcrowding and dissatisfaction in these areas have severe, often undiagnosed and untreated mental health implications, contributing to feelings of hopelessness and learned helplessness (Weimann and Oni 2019).

6.5 Methodology

In February 2022, a qualitative case study approach was used to understand the experiences and perspectives of residents and municipal officials in the context of liveability and urban health. The study involved participants of different ages, genders, physical abilities, employment status, and length of residence in the dormitory, ensuring diverse perspectives were obtained. Participants engaged in structured discussions, with topic guides containing open-ended questions and prompts prepared in advance. The discussions focused on residents' experiences and perceptions regarding housing, service delivery, and their impact on health, as well as strategies for improving liveability and urban health. Ethical considerations were followed, ensuring informed consent, confidentiality, and respect for participants' rights.

In-depth interviews were conducted with seven municipal officials in their respective offices or via WhatsApp video, and sixteen residents at the hostel to explore their knowledge and experiences regarding housing, infrastructure, access to services, and governance issues. Two community meetings involving Hostel 1 residents and community leaders were held to foster a participatory approach and gain insight into lived experiences. Collective discussions, knowledge sharing, and collaborative problem-solving were facilitated to understand the community's priorities, concerns, and aspirations related to housing, infrastructure, and service provision. A focus group discussion with five members of the community's steering committee was also held to gain a collective understanding of the linkages between SDG 11 and urban health in the context of Hostel 1.

The discussions were conducted with mutual respect and active listening, allowing for open dialogue, comfortable interaction, knowledge sharing, and exploration of themes related to the research topic. Detailed notes and audio recordings were taken during meetings, capturing key themes, ideas, non-verbal cues, group dynamics, and contextual observations for analysis. Data collected were transcribed and thematically analysed using NVivo 14 to identify and interpret patterns related to SDG 11, liveability, and urban health. Both inductive and deductive approaches were applied to identify emerging themes which included housing adequacy, infrastructure, service delivery, safety, collaborative approaches, and governance.

6.6 Findings Linking Liveability to SDG11 and Urban Health at Hostel 1

6.6.1 Background

According to participants, White City, popularly known as Hostel 1, was established as a male-only residence in Bochabela, Bloemfontein around 1945–1950. It consists of some communal dwellings and is situated approximately 5 km from Bloemfontein CBD, across Dr. Belcher Road (leading to Eastern Cape and Lesotho) and Mkhuhlane Street (the high street into Bochabela and other Mangaung Township neighbourhoods) (Google 2024). At the time of the study, it was difficult to determine the exact number of rooms or families at Hostel 1 due to the complex allocation of space. The data indicated that most residents were Sotho people from various regions of the Free State province, Eastern Cape and Lesotho. There were also some Tswana, Xhosa, and Zulu people from KwaZulu Natal and the Western Cape. The participants were born between 1971–2003 and had lived in Hostel 1 for periods ranging from 5 to 45 years. Those who stayed for up to five years came to Bloemfontein for employment opportunities. Their main reasons for choosing Hostel 1 were affordability and proximity to socio-economic opportunities. Those who had stayed longer (e.g., 16–45 years) were either born in the area, inherited their rooms from their parents, or moved closer to family. The participants' occupations included formal business (construction work, plumbing, security, retail and machine operation), government work (general positions), household-based work (domestic roles), and informal activities (general and construction). Only two were unemployed, with one being a student and the other searching for a job.

6.6.2 The Nature of Housing Adequacy in Hostel 1

This theme was based on tenure security, physical conditions, plans and safety variables. The topic of tenure was difficult as the participants were hesitant to talk about it. This was probably caused by the contested views of the hostel ownership and apparent uncertainty on whether the municipality would allow their tenure indefinitely. During apartheid, Hostel 1 served as dormitory accommodation for Black male migrant workers. However, following democracy, the composition of its residents changed significantly. Family members of the male workers began moving into the rooms, seeking socio-economic opportunities. While the governance transition occurred, the occupants assumed unofficial ownership, which was subsequently passed down to their children. As highlighted, *"We got this place from our parents. This place is our birthright. We were born and raised here"* (Resident N from Hostel 1, Bloemfontein, 2022, verbal communication). Most Hostel 1 residents inherited their rooms from previous generations, fostering a sense of attachment to the place. Thus, they did not plan to sell their units but to renovate them when finances permitted, and then pass them over to their children. However, they said that their lack of formal proof of ownership like title deeds, caused considerable stress and uncertainty. Resident H (Resident H from Hostel 1, Bloemfontein, 2022, verbal communication) said of this, *"They [municipality] can kick us out at any moment."* Those who readily admitted that the hostel was government property planned to relocate eventually: *"It is a government house, so I do not have plans for it"* (Resident C). Overall, the popular finding was that though the residents had no title deeds, they perceived Hostel 1 as their fathers' land, which made it theirs.

To cope with the increased demand for housing, the residents converted the ablution blocks and hostel hall into residential rooms, using partition boards. Furthermore, to accommodate the growing number of families, some rooms were

Fig. 6.1 (**a**) Hostel 1 external rooms; (**b**) internal rooms. (Source: Authors 2022)

partitioned using curtains or zinc sheets for sharing between up to four or more families. This added to the congestion of the residence. Other rooms were rented to migrants to generate income for the "owners", who either moved elsewhere or shared lodgings with the tenants. The buildings have internal and external entries as highlighted in Fig. 6.1.

Presently, the rooms in Hostel 1 are in a state of disrepair, evident through dilapidated walls, roofing, doors, and windows. These structural deficiencies give rise to safety and security challenges. In times of rain and strong winds, water inundates the rooms, leading to roofing damage and mould on the walls, which were reported to contribute to distress, flu and chest or breathing problems. Moreover, makeshift doors and mostly broken windows serve as easy entry points for criminals, resulting in frequent break-ins, attacks, theft, and incidents of sexual assault. Many residents claimed to feel constantly unsafe in the hostel.

The change in occupancy and increased number of residents in the hostel brought cultural changes, especially in terms of spoken languages, behaviours, social relations and the scope of economic opportunities. For example, informal structures made of zinc sheets have been set up within the compound, though some residents complain that this was done by "outsiders." The outsiders were either South African citizens from other parts of Bloemfontein or South Africa, and foreigners. These outsiders are perceived as troublesome and criminal, especially foreigners. They were said to "promote crime, sell drugs and commit murder." They were also blamed for chaotic behaviour and gangsterism. Moreover, those who considered themselves native to the hostel were concerned that the outsiders took housing and job opportunities from them. All these perceptions contributed to bad relations between these two groups.

6.6.3 Infrastructure and Service Delivery

The influx of residents and the transformation of the hostel into a family setting put considerable strain on the existing services and infrastructure, which were originally designed to cater to single male workers. The transport system was efficient and easily accessible with the roads in acceptable condition. The findings indicate that it was not common to have road accidents or transport-related problems in the area, except for having to walk in the mud on rainy days when they went to the bus stop. There was ongoing road maintenance and construction during the study, which was part of the municipality's rapid bus transit initiative. The residents were also satisfied with health services as the public Pelonomi Hospital was within a short walking distance of the hostel.

Fig. 6.2 The dumping site outside one of Hostel 1's buildings. (Source: Authors 2022)

However, the government's inadequate upkeep and inefficient services were a source of dissatisfaction among the residents. The most pressing challenge faced by the residents was the absence of toilet facilities for over two years, resulting in significant health and safety concerns, particularly for women, children, and individuals with physical disabilities. Residents resort to open spaces or use buckets, creating discomfort and unsanitary conditions around the buildings. As emphasized by a physically challenged Resident J (from Hostel 1, Bloemfontein, 2022, verbal communication) *"Look where I am sitting (pointing at the wheelchair), they say we must go relieve ourselves in the wild. How do I do that?"* The open space, which serves as an area for physical activities like soccer matches, also suffers from illegal dumping due to inefficient municipal garbage collection (Fig. 6.2). Additionally, some illegal houses (shacks) have been constructed in the open space, though residents claim to continuously hinder further efforts by illegal settlers. Despite being used as a pathway, the open space is considered unsafe, particularly at night, due to its neglected condition and lack of lighting. Residents mentioned instances of attacks and robbery here.

The environment is generally considered unsafe for children, as they play close to dumping sites, unsupervised buildings and an unkempt open space (Fig. 6.3). These have become prevalent issues of concern among the residents who are dissatisfied with the congested and unsafe living conditions within the buildings. Some mentioned that the children had skin problems and often fell sick. They also felt that such conditions affected the mental well-being of residents.

Residents' attempts to seek government support for constructing toilets were hampered by their insecure tenure, preventing them from building their facilities. The area once had communal ablution blocks, which have been converted into residential rooms. Temporary toilets supplied by the municipality were removed due

Fig. 6.3 Children playing close to a dumping site with a non-functional sewer and tap in view. (Source: Authors 2022)

to their high upkeep costs. However, residents suspect that this removal was a form of punishment, as they are often considered illegal settlers who are undeserving of services, except during election times when politicians solicit their votes. The option of pit latrines is not favoured by the residents, as they perceive them as dangerous, especially for children. Similarly, communal toilets were also rejected due to safety concerns, particularly at night. Specifically, there is a fear and suspicion towards foreigners, whom residents believe escalate crime, including murder. To address the toilet issue, residents proposed using existing running water and sewer systems to construct new toilets in each block, leveraging the available infrastructure to improve sanitation conditions within the hostel.

6.6.4 Governance Issues and Collaborative Efforts

Access to the Hostel was facilitated through gatekeepers represented by the Mangaung Service Delivery Forum (MSDF), an autonomous community committee not aligned with any political party. The MSDF functioned as an intermediary between the community and various levels of government, advocating and negotiating for improved service delivery to address the grievances of the residents. The escalating situation at the hostel prompted the MSDF to organize multiple meetings between the residents and local authorities (Fig. 6.4). One of these meetings was attended by the MMC of Housing who sought to understand the community's concerns and assess the living conditions. During the meeting, the authorities promised to promptly set up temporary toilets as an urgent measure while finding a more permanent solution.

However, as time passed, it appeared that the authorities were employing delay tactics and did not fulfil their commitments. The community noticed a lack of progress in addressing their grievances. The area councillor was perceived to be ignoring their communication, while also giving inaccurate reports to the municipality. In addition to the toilet issue, the committee also approached the police authorities to address the safety concerns faced by the residents, particularly women and children. According to the residents, the government and the private sector should intervene by addressing housing provision, job creation, community support, and engagement, as well as promoting good governance. Additionally, they emphasized the impor-

Fig. 6.4 Community meeting at Hostel 1. (Source: Authors 2022)

tance of community unity and collaboration to tackle issues such as illegal dumping, neighbourhood cleanup, and devising collective solutions. They also suggested collaborating to improve housing conditions and establishing youth centres for technical skills development.

In trying to understand why the hostel existed in its present state, government officials were engaged. Their feedback indicated that the hostel was supposed to have been demolished, as new social housing was constructed to accommodate the Hostel 1 residents. However, the committee indicated that they were still negotiating with the government to get new houses. As stipulated by a member: *"They built those houses you see there across, they were supposed to be ours. But to our surprise, we saw people we don't know where they come from moving in there. Only a few of us got the houses, not all of us."* There was still hope among some residents that the promised houses would be allocated to them: *"We plan to move to government housing that was promised"* (Resident F, Member of community committee at Hostel 1, Bloemfontein, 2022, verbal communication). The social housing was supposed to be developed in three phases to replace the hostel and enable liveability around the area. The residents claimed that when the first phase was completed, the then-existing hostel toilets were demolished, yet *"officials gave the new houses to their people"* leaving residents yearning for a place of belonging. The committee likened this to *"accumulation by dispossession"* as outsiders who gained entry to the new houses did so at the expense of the hostel residents.

To address the question of incomplete social housing phases and questionable allocation, officials highlighted that while regulations were essential for housing, job creation, and economic development, their effectiveness was hindered by weak applicability. Challenges in implementation arise from inconsistencies among relevant departments and stakeholders, particularly for townships. Intergovernmental hindrances, such as red tape, power relations, and political interference, also contribute to institutional challenges. Planning decisions related to service delivery and maintenance are delayed due to interactions between the three spheres of government. Moreover, inadequate budgeting for operational and maintenance costs, as well as new projects, led to ineffective service delivery, public contestations, and a significant debt in levies owed. Officials expressed reluctance to share information and ideas due to job security concerns, possibly related to political interference and the need for professional bodies to intervene. Additionally, councillors avoided conveying unfavourable news to maintain popularity, resulting in inefficient communication with the public. These findings highlighted that effective communication is crucial for stakeholder engagement and collective problem-solving. Table 6.5 summarizes the findings of this study against liveability, SDG 11 and urban health.

6.7 Discussion and Policy Implications

As cities continue to grow as economic hubs, rural-urban migration drives population growth, necessitating government prioritization of economic growth while creating liveable, safe, and resilient communities for residents to enjoy a good quality of life. Urbanization significantly affects population health in urban areas (Thomas et al. 2016). As urbanization and governance changes occurred in Bloemfontein, Hostel 1 residents adapted to their changing needs and circumstances, with the outcomes highlighting the challenges posed by the mismatch between the original design of the hostel and its evolving use. The residents of Hostel 1 face poor housing conditions and inadequate access to basic services like sanitation and security, leading to overcrowding and the spread of diseases including diarrhoea, stomach pains, cholera, hepatitis E and "women issues" (Resident X 2022) due to poor waste management. In spaces with poor sanitation and hygiene, air quality also contributes to urban health implications with at least 91% of people breathing polluted air (WHO 2021).

A dignified life is essential for everyone, including Hostel 1 residents, making health a priority in urban sustainable development (Rebecchi and Capolongo 2021). The SDG 11 aims to create sustainable cities with a focus on urban health aspects such as clean water and sanitation. This requires strengthening capacities for urban planning and enhancing livelihoods. Proactive urban policies are needed to address the critical challenges of the city and involve all sectors—public, private, civil society, academia, and the community—to achieve SDG 11 and overall urban health. It is vital that the opportunities presented by urbanization be optimized to deal with urban health issues such as those presented in Hotel 1. The localization of SDG 11 implementation is pivotal and requires collaborative action from all sectors.

According to Carmichael et al. (2019), the places we inhabit profoundly impact our physical, mental, social, environmental, and economic well-being. Poor environmental quality exposes us to risk factors like urban heat islands, pollution, and traffic-related issues (Rebecchi and Capolongo 2021). Proper water management is crucial for urban areas to ensure safe drinking water and sanitation for all, as inadequate water management can lead to diseases like cholera and hepatitis B. Sustainable water management approaches, such as wastewater treatment and improving water sources, are essential due to increasing demands. It should be noted that people residing in overcrowded spaces rarely prioritize their health and safety until an epidemic erupts. Thus, it rests on the government and urban planners to ensure that all spaces/neighbourhoods are liveable, safe and resilient.

Urban planners play a significant role in enabling urban health, considering that it relies not only on healthcare services but also on the quality of natural and built environments (Rebecchi and Capolongo 2021). The SDG 11 targets access to safe, adequate, and affordable housing, emphasizing the importance of inclusive policies and participatory policymaking, which address unplanned urbanization. Carmichael et al. (2019) stress that policies regulating land use, connectivity, density, transport, and green infrastructure can lead to improved health outcomes, highlighting the role of spatial planning in supporting health. Promoting urban health can also contribute to thriving housing markets and local economic growth.

To improve urban health and address public health challenges, urban policies should integrate

Table 6.5 Sustainable Development Goal (SDG) 11 and urban health experiences in Hostel 1

Liveability Requirement	SDG 11 target	Urban health directive	Hostel 1 experience
Sociological + functional	11.1 Safe and affordable housing	Physical and mental health e.g., respiratory conditions, heat shocks, injuries, psychological distress, and child development	Dilapidated, inadequate, congested dormitories without proper sanitation and water, lighting and air circulation. During the COVID-19 pandemic, social distancing was difficult. There is a lack of privacy and safety
Functional + social (mental and physical well-being)	11.2 Affordable and sustainable transport systems	Road accidents	Efficient and adequate transport system. Also walkable to central business district and other areas of opportunity e.g., industrial and commercial
Cultural determinant (identity and belonging)	11.3 Inclusive and sustainable urbanization	Social disruptions that may lead to injury, death, stress	Migrants from other provinces and countries caused tension as they were perceived to infiltrate the contested space (Hostel 1), decrease job and business opportunities, and escalate crime including murder. Relations with locals were tense due to the scarcity of resources and fear
Cultural determinant (identity and belonging)	11. 4 Protect the world's culture and natural heritage	Place attachment and feeling of belonging and responsibility contributes to mental well-being	Most residents inherited their rooms from forebears. There was sense of attachment but stress due to lack of formal proof of ownership and lack of upkeep. Relations with migrants were bad
Environmental protection	11.5 Reduce the adverse effects of natural disasters	Flooding and strong winds pose physical and mental health challenges	The infrastructure is aged and risky during environmental disasters such as storms, strong winds and flooding
Environmental protection	11.6 Reduce the environmental impact of cities	Air, water, and land pollution contribute to health risks including communicable diseases	Inefficient waste management, with alternative solid and liquid waste disposal causing land and air pollution and posing danger to residents, especially children, women, disabled
Functional	11.7 Provide access to safe and inclusive green and public spaces	Mental well-being of urban residents and physical health through exercise and spending time in clean environment	Public space unkempt, used as an alternative dumping site and for informal settlers. No nighttime lighting and safety a challenge as some robberies and attacks occurred

(continued)

Table 6.5 (continued)

Liveability Requirement	SDG 11 target	Urban health directive	Hostel 1 experience
Functional (governance)	11.A. Strong national and regional development planning	Availability of necessary infrastructure and social services for healthy cities	Integrated development plan exists, the issue is inefficient public participation and resources. Residents have feelings of unsafety, exclusion, injustice, demoralization. Municipality and local authorities tread with caution in improving services and infrastructure

Source: Authors (2023)

public health considerations at multiple levels (Carmichael et al. 2019). Multidisciplinary urban development should incorporate tools for capacity building, research, and strengthening urban governance for health. Governments and think tanks should promote urban health research, planning, and implementation to understand the roles of SDG 11 in achieving urban health (Thomas et al. 2016). Policies aimed at improving urban health should clearly define stakeholders' roles to bridge the gap between urban development and health implementation. Sustainable urban areas with proper planning have better access to clean water and sanitation, contributing to residents' overall health. Urban planning should conduct health-inclusive environmental assessments to understand the health implications of proposed developments. Achieving urban health involves making cities sustainable and ensuring access to clean water and sanitation. The provision of safe and affordable housing ties in with the enhancement of the social and economic well-being of the community of Hostel 1.

6.8 Conclusion and Recommendations

In conclusion, this chapter highlights that the liveability status of Hostel 1 residents is severely compromised due to a range of challenges. The housing conditions are in a state of disrepair, with dilapidated walls, roofing, doors, and windows, posing significant safety and security risks. These structural deficiencies contribute to health and safety concerns, including mould formation, respiratory issues, and incidents of criminal activity such as break-ins and theft. The absence of formal proof of ownership has led to stress and uncertainty among residents, while the lack of adequate sanitation facilities, including the absence of toilet facilities, further exacerbates health and safety risks, particularly for vulnerable groups such as women, children, and individuals with physical disabilities. The open space in the neighbourhood is marred by illegal dumping and the presence of illegal houses, contributing to safety concerns, especially at night due to inadequate lighting. These challenges underscore the compromised liveability status of Hostel 1 residents, emphasizing the urgent need for sustainable urban development to improve their living conditions and overall well-being. To improve urban planning and policies to address the health and safety challenges faced by the residents of Hostel 1 and achieve SDG 11 objectives, several key strategies can be implemented:

1. **Inclusive and sustainable urbanization:** Involve the residents in decision-making and prioritize their needs in urban development plans for inclusive and sustainable urbanization.
2. **Affordable and sustainable housing:** Ensure access to adequate, safe, and affordable housing and basic services through upgrading Hostel 1 and providing affordable housing options for residents.

3. **Environmental impact reduction:** Efforts must be taken to reduce the environmental impact of the hostel on environmental quality and waste management. This includes improving waste management practices, reducing pollution, and promoting sustainable environmental practices among residents.
4. **Access to safe and inclusive green spaces:** Create greening of the open space and maintain its safety, inclusivity, and accessibility for all community members. Strong national and regional development planning; strengthen national and regional development planning to support positive links between urban, peri-urban, and rural areas, ensuring balanced territory development.
5. **Disaster risk reduction:** Implement policies and plans for disaster risk reduction by integrating related strategies into Hostel 1 planning and development initiatives.
6. **Collaboration and community engagement:** Encourage collaboration and community engagement in urban planning and policy-making processes by fostering partnerships between government, community organizations, private sector, academia, and other stakeholders to address the specific needs and challenges faced by the residents of Hostel 1.

By implementing these strategies and integrating SDG 11 principles into urban planning, it is possible to create healthier, more sustainable, and inclusive urban environments. This will contribute to achieving SDG 11 objectives and address the health and safety challenges faced by marginalized populations.

References

Abdelaty H, Weiss D, Mangelkramer D (2023) Climate policy in developing countries: analysis of climate mitigation and adaptation measures in Egypt. Sustain For 15(11):9121

Agboola O, Babatunde DE, Fayomi OS, Sadiku ER, Popoola P, Moropeng L et al (2020) A review on the impact of mining operation: monitoring, assessment and management. Results Eng 8:100181

Ban KM (2016) Sustainable development goals. News Surv 37(02):18–19

Beck EJ, Shields JM, Tanna G, Henning G, De Vega I, Andrews G et al (2018) Developing and implementing national health identifiers in resource limited countries: why, what, who, when and how? Glob Health Action 11(1):1440782

Boyce MR, Katz R, Standley CJ (2019) Risk factors for infectious diseases in urban environments of sub-saharan Africa: A systematic review and critical appraisal of evidence. Trop Med Infect Dis 4(4):123. https://doi.org/10.3390/tropicalmed4040123

Budiarti PW, Fathani TF, Faris F (2020) Analysis of rainfall-triggered landslide in Baleagung Village, Magelang Regency, Central Java. In E3S Web of Conferences. EDP Sciences, 200:01001

Carmichael L, Townshend TG, Fischer TB, Lock K, Petrokofsky C, Sheppard A, Sweeting D, Ogalvie F (2019) Urban planning as an enabler of urban health: challenges and good practice in England following the 2012 planning and public health reforms. Land Use Policy 84:154–162

Chopde NR, Nichat M (2013) Landmark based shortest path detection by using A* and Haversine formula. Int J Innov Res Comput Comm Eng 1(2):298–302

Chronaki CE, Kontoyiannis V, Panagopoulos D, Katehakis DG, Vourvahakis D, Koutentaki-Mountraki K (2011) Interoperability in disaster medicine and emergency management. J Health Inform 3

Elbeltagi E, Hegazy T, Eldosouky A (2004) Dynamic layout of construction temporary facilities considering safety. J Constr Eng Manag 130(4):534–541

Fathani TF, Karnawati D, Wilopo W (2016) An integrated methodology to develop a standard for landslide early warning systems. Nat Hazards Earth Syst Sci 16(9):2123–2135

Fathani TF, Karnawati D, Wilopo W, Setiawan H (2023) Strengthening the resilience by implementing a standard for landslide early warning system. Prog Landslide Res Technol 1(1):277–284. Cham: Springer International Publishing

Fisonga M, Wang F, Mutambo V (2019) Sustainable utilization of copper tailings and tyre-derived aggregates in highway concrete traffic barriers. Constr Build Mater 216:29–39

Galea S, Vlahov D (2005) Urban health: evidence, challenges, and directions. Annu Rev Public Health 26(1):341–365. https://doi.org/10.1146/annurev.publhealth.26.021304.144708

Gambe TR, Turok I, Visagie J (2023) The trajectories of urbanisation in southern Africa: a comparative analysis. Habitat Int 132:102747. https://doi.org/10.1016/j.habitatint.2023.102747

Gamperl M, Singer J, Garcia-Londoño C, Seiler L, Castañeda J, Cerón-Hernandez D, Thuro K (2023) Recommendations for landslide early warning systems in informal settlements based on a case study in Medellín, Colombia. Land 12(7):1451

Google (2024). https://www.google.com/maps/dir/Bloemfontein/White+City,+Mangaung/@-29.1341417,26.2315426,1448m/data=!3m1!1e3!4m14!4m13!1m5!1m1!1s0x1e8fc57b9f07bf8 b:0x1f16f55bb211b749!2m2!1d26.214937!2d-29.1129237!1m5!1m1!1s0x1e8fcf78dacecb95:0xd9424bd4f15bd551!2m2!1d26.2371061!2d-29.1345049!3e0?entry=ttu

Gumiran BA, Daag A (2021) Negotiated participatory action research for multi-stakeholder implementation of early warning systems for landslides. Int J Disast Risk Red 58:102184

Hajduk S (2020) Using multivariate statistical methods to assess the urban smartness on the example of selected European cities. PLoS One 15(12):e0240260

Hakim S, Mohsen A (2017) Work-related and ergonomic risk factors associated with low back pain among bus drivers. J Egypt Public Health Assoc 92(3):195–201

Huang B, Gao X, Xu X, Song J, Geng Y, Sarkis J et al (2020) A life cycle thinking framework to mitigate the environmental impact of building materials. One Earth 3(5):564–573

Haywood LK, Kapwata T, Oelofse S, Breetzke G, Wright CY (2021) Waste disposal practices in Low-Income Settlements of South Africa. Int J Environ Res Public Health 18(15):8176. https://doi.org/10.3390/ijerph18158176

ISO 22327 (2018) Security and resilience—emergency management—guidelines for implementation of a community-based landslide early warning system. Int Org Standard, Switzerland

King R, Orloff M, Virsilas T, Pande T (2017) Confronting the urban housing crisis in the global south: adequate, secure, and affordable housing. World Resources Institute, Washington, DC

Lawanson T, Salau T, Yadua O (2013) Conceptualizing the liveable African city. J Const Project Manag Innov 3(1):573–588. https://journals.co.za/doi/pdf/10.10520/EJC136878

Lemon A (1991) The apartheid city. In: Lemon A (ed) Homes apart: South Africa's segregated cities. Indiana University Press, Cape Town, pp 1–25

Marais L, Cloete J (2014) "Dying to get a house?" The health outcomes of the South African low-income housing programme. Habitat International 43:48–60

Matamanda AR, Nel V, Dunn M, Mgwele A, Rammile S, Leboto-Khetsi L, Kohima J, Ngo P (2022) Sustainable development goals and urban health challenges in informal settlements of Mangaung Municipality, South Africa. In: Filho WL et al (eds) Handbook of sustainability science in the future. Springer Nature Switzerland AG. https://doi.org/10.1007/978-3-030-68074-9_157-1

Newton C, Schuermans N (2013) More than twenty years after the repeal of the Group Areas Act: housing, spatial planning and urban development in post-apartheid South Africa. J Hous Built Environ 28:579–587

Omary Z, Lupiana D, Mtenzi F, Wu B (2009) Challenges to E-healthcare adoption in developing countries: a case study of Tanzania. In: 2009 first international conference on networked digital technologies 201–209. IEEE

O'Neill A (2024) South Africa: Urbanization from 2013-2023. https://www.statista.com/statistics/455931/urbanization-in-south-africa/

Opoko AP, Oluwatayo A (2014) Trends in urbanisation: implication for planning and low-income housing delivery in Lagos, Nigeria. Archit Res 4(1A):15–26

Our World in Data (2023) Make cities inclusive, safe, resilient and sustainable. https://ourworldindata.org/sdgs/sustainable-cities

Paul A, Sen J (2020) A critical review of liveability approaches and their dimensions. Geoforum 117:90–92. https://doi.org/10.1016/j.geoforum.2020.09.008

Pecorella T, Ronga LS, Chiti F, Jayousi S, Franck L (2015) Emergency satellite communications: research and standardization activities. IEEE Commun Mag 53(5):170–177

Peters LE, Clark-Ginsberg A, McCaul B, Cáceres G, Nuñez AL, Balagna J, Van Den Hoek J (2022) Informality, violence, and disaster risks: coproducing inclusive early warning and response systems in urban informal settlements in Honduras. Front Clim 4:937244

Polasi L, Matinise S, Oelofse S (2020) South African Municipal Waste Management Systems: Challenges and Solutions. https://www.researchgate.net/publication/346025379_SOUTH_AFRICAN_MUNICIPAL_WASTE_MANAGEMENT_SYSTEMS_CHALLENGES_AND_SOLUTIONS

Rebecchi A, Capolongo S (2021) Healthy design and urban planning strategies framing the SDG-11 sustainable cities and communities. 14th European Public Health Conference

Sachs JD, Kroll C, Lafortune G, Fuller G, Woelm F (2022) Country Profiles. In: Sustainable Development Report 2022. Cambridge University Press 2022:83–83

Sarkarb A, Bardhana R (2020) Socio-physical liveability through socio-spatiality in low-income resettlement archetypes – a case of slum rehabilitation housing in Mumbai, India. Cities 105:102840. https://doi.org/10.1016/j.cities.2020.102840

Sekoboto L, Landman K (2018) Searching for more than just a house? The extent to which government provided RDP housing compared to self-help housing empower poor communities in South Africa. Urban Forum 30:97–113

Skalickya V, Čerpesb I (2019) Comprehensive assessment methodology for liveable residential environment. Cities 94:44–54. https://doi.org/10.1016/j.cities.2019.05.020

Sochacka BA, Kenway SJ, Renouf MA (2021) Liveability and its interpretation in urban water management: systematic literature review. Cities 113:103154. https://doi.org/10.1016/j.cities.2021.103154

Statistics South Africa (2019) Sustainable development goals country report South Africa. https://www.statssa.gov.za/MDG/SDGs_Country_Report_2019_South_Africa.pdf

Statistics South Africa (2023) Sustainable development goals country report South Africa. https://www.statssa.gov.za/MDG/SDG_Country_report.pdf

Strauss M (2019) A historical exposition of spatial injustice and segregated urban settlement in South Africa. Fundamina 25(2):135–168. https://doi.org/10.17159/2411-7870/2019/v25n2a6

Szpak A (2016) Right to the city as a new paradigm within the concept of human security. Polish Polit Sci Yearb 45(1):367–381

Takiya H, Negreiros I, Yamamura CLK, Quintanilha JA, Machado CAS, Abiko A et al (2022) Application of open government data to sustainable city indicators: a megacity case study. Sustain For 14(14):8802

Tanwar S, Tyagi S, Kumar S (2018) The role of internet of things and smart grid for the development of a smart city. In: Intelligent communication and computational technologies: proceedings of internet of things for technological development, IoT4TD 2017:23–33. Springer Singapore

Thomas YF, Boufford JI, Talukder SH (2016) Focusing on health to advance sustainable urban transitions. J Urban Health Bull New York Acad Med 93(1):1–5. https://doi.org/10.1007/s11524-016-0037-x

Thompson AA, Matamale L, Kharidza SD (2012) Impact of climate change on children's health in Limpopo Province, South Africa. Int J Environ Res Public Health 9(3):831–854. https://doi.org/10.3390/ijerph9030831

United Nations (2015) Transforming our world: the 2030 agenda for sustainable development (A/RES/70/1). https://sustainabledevelopment.un.org/content/documents/21252030%20Agenda%20for%20Sustainable%20Development%20web.pdf

Weimann A, Oni T (2019) A systematized review of the health impacts of urban informal settlements and implications for upgrading interventions in South Africa, a rapidly urbanizing middle-income country. Int J Environ Res Public Health 16:3608. https://doi.org/10.3390/ijerph16193608

World Health Organization (2016) Urban health. https://www.who.int/westernpacific/health-topics/urban-health

World Health Organization (2021) Urban Health. https://www.who.int/news-room/fact-sheets/details/urban-health

Wuerzer T (2014) Urban health. In: Michalos AC (ed) Encyclopedia of quality of life and well-being research. Springer, Dordrecht. https://doi.org/10.1007/978-94-007-0753-5_3127

Zimring CA, Rathje WL (eds) (2012) Encyclopedia of consumption and waste: the social science of garbage, vol 1. Sage

7 Informal Settlements Under the Impact of Climate Change and the Community Health Factor in Mangaung Metropolitan Municipality, South Africa

Halleluah Chirisa, Adriaan Van Der Walt, and Abraham R. Matamanda

7.1 Introduction

The continued prevalence of extreme climatic events (for example, heatwaves, cold events, floods, and droughts) and other annual changes in weather conditions have negative impacts on humans (Zhang et al. 2019). The increasing occurrence of extreme climatic events implies that governments and local authorities, with local communities, must rethink housing provisions for low-income earners living in informal settlements. In recent years South Africa has experienced episodic extreme climatic events including heavy rainfall and flooding especially Western Cape and KwaZulu Natal Provinces. For example, during the second week of September 2023, the Western Cape received more than two months' worth of rainfall attributed to climate change (van Diemen 2023). Informal settlements emerge as one of the vulnerable spaces to climate change disasters. Specifically, the location of the informal settlements in low-lying areas has worsened the situation and increased the impacts of flood risks for some poor communities. This was experienced in Cape Town during the June 2023 floods (Isaacs 2023).

Climate change results from a continuous increase in the emission of greenhouse gases and other climate forces (Dino and Akgül 2019). The potential consequences of climate change include negative impacts on human health (Zhao et al. 2022). People living in informal settlements are more vulnerable to the health effects of climate change due to the absence of services such as water and sanitation, poor housing and lack of healthcare facilities (Satterthwaite et al. 2020; Corburn and Sverdlik 2019). Residents are likely to face high risks as they lack the infrastructure required to withstand and cope with extreme weather conditions such as floods. Building structures in informal settlements increases residents' susceptibility to climate change.

This background on climate change and informal settlement brings to attention urban health challenges within the context of the Sustainable Development Goals (SDGs). Specifically, the chapter relates to SDG 13 on climate action which aims to "strengthen resilience and adaptive capacity to climate-related hazards and natural disasters in all countries". The aim of SDG 13 is to explore the nexus of climate change extremes and urban health among poor neighbourhoods in Mangaung Metro Municipality, South Africa.

H. Chirisa (✉) · A. Van Der Walt
Department of Geography, University of the Free State, Bloemfontein, South Africa

A. R. Matamanda
Department of Urban & Regional Planning, University of the Free State, Bloemfontein, South Africa

A. R. Matamanda, V. Nel (eds.), *Sustainable Development Goals and Urban Health*, Sustainable Development Goals Series, https://doi.org/10.1007/978-3-031-68734-1_7

Urban health and related diseases that emanate from climate change include heat stress, waterborne diseases, and increased prevalence of vectors, viruses and airborne diseases (Linares et al. 2020). Climate change and environmental degradation are equity issues that undermine the rights of every citizen, especially the most disadvantaged (Hoeltl et al. 2020).

7.2 Literature Review

In recent years, much has been written about housing, health and climate change, as disparate but related subjects. Missing in the literature and practical discourse in South Africa is a nuanced discussion around informal settlements and housing and the impacts of climate change. Five SDG 13 targets and indicators on climate are clear and are summarised in Table 7.1. Climate action focuses on creating resilient communities that can adapt to climate change. This is linked to the issue of adequate and affordable housing and infrastructure that can withstand the changing climate and related extremes. Adequate, liveable and well-serviced housing leads benefits in health, education and economic opportunities (Badland and Pearce 2019).

Table 7.1 Sustainable Development Goal (SDG) 13 targets and indicators (adopted from Lee et al. 2023)

Target	Description
Target 13.1	Strengthen resilience and adaptive capacity to climate-related hazards and natural disasters in all countries
Target 13.2	Integrate climate change measures into national policies, strategies and planning
Target 13.3	Improve education, awareness-raising and human and institutional capacity on climate change mitigation, adaptation, impact reduction and early warning
Target 13.a	Implement the commitment undertaken by developed-country parties to the United Nations framework convention on climate change to a goal of mobilizing jointly $100 billion annually by 2020 from all sources to address the needs of developing countries in the context of meaningful mitigation actions and transparency on implementation and fully operationalize the green climate fund through its capitalization as soon as possible
Target 13.b	Promote mechanisms for raising capacity for effective climate change-related planning and management in least developed countries and small island developing states, including focusing on women, youth and local and marginalized communities

The damage caused by extreme climate events and related health impacts calls for housing improvements, especially in marginal spaces (Gajjar et al. 2019; Madhanagopal and Jacob 2022). Housing improvement is often a ladder out of poverty for families. These changes benefit the larger community, reduce inequalities and build resilience against economic and natural disasters, even for the poor who are the majority living in informal settlements.

Table 7.1 summarises SDG 13, showing the targets, indicators, and the monitoring mechanisms. The main thrust of SDG 13 is on climate action. Climate action focuses on the following targets: building resilient communities that can adapt to climate disasters and extremes; national policies that are responsive to climate change, educating communities on mitigation, adaptation, impact reduction and early warning systems; and fundraising and commitment by developed nations (Lee et al. 2023). It also focuses on building mechanisms that enable developing countries to cope with climate change-related disasters.

Climate change impacts communities differently, with the least resilient communities, countries and regions carrying the highest burden. Informal settlements require strategies that increase resilience and adaptation capacity to cope with climate change. There is an intricate relationship between SDG 13 on climate action, SDG 3 on good health and SDG 11 on sustainable cities. In relating the issue of resilience and adaptation to climate change within the informal settlements, one would need to unpack SDG 3. Due to rapid urbanisation in developing countries, urban areas, especially in the Global South, are characterised by informal settlements. The large influx of people through rural-urban migration and natural increase through excess births over deaths in urban areas are putting pressure on the available resources, especially housing and sanitation. With the increasing extreme climate events, climate change-related hazards have also increased.

Climate change and related losses and damages are unequally distributed across systems, regions and sectors. Economic damages from climate change are found in climate-exposed sectors, such as agriculture, forestry, fishery, energy and tourism, the economies of sub-Saharan African countries and the Pacific Islands are affected because they rely on natural resources (Ibe and Amikuzuno 2019). Individual livelihoods are affected by the destruction of homes and infrastructure, loss of property and income, human health and food security, with adverse effects on gender and social equity (Matamanda and Nel 2021).

In urban areas, climate change caused adverse impacts on human health, livelihoods and key infrastructure. Hot extremes have intensified in cities (Mentaschi et al. 2022). Urban infrastructure, including transportation, water, sanitation, and energy systems are compromised by extreme and slow-onset events that result in economic losses, disruptions of services and negative impacts on the general well-being of people especially those residing in informal settlements (Weimann and Oni 2019).

The effects of extreme rains were experienced in Durban (South Africa) in 2022. Subsequent flooding causes direct loss of life through drowning and diarrhoeal infections due to excreta mixing with drinking water in informal settlements. During 11–12 April 2022, South Africa experienced extreme rainfall equivalent to a year's worth rainfall on the eastern coast. About 400 people died, 1, 200 households were destroyed and 40, 000 people were relocated due to landslides and floods (Tandon 2022). This extreme event was declared a national disaster with R1.52 billion in structural damages. Due to their location and poor building structures, informal settlements were affected the most.

7.3 National Context, Trends and Scholarship

South African cities are hugely inadequate shaped by discriminatory and repressive apartheid planning and further expanded by powerful and far-from-equitable market processes, driving apartheid's planned inequality and exclusion even deeper, and effortlessly overriding attempts at urban democratisation and integration. Informal settlements are a major challenge in the urban landscape, exacerbating issues related to poverty, inadequate infrastructure, housing and poor living conditions. The effects of climate change, poverty and informality are strongly intertwined and meet at the urban level. However, this is not fully reflected in international climate action and poverty reduction efforts. There is a continued proliferation of informal settlements despite laws being implemented to curb the growth of slums.

7.3.1 Community Health

Communities suffer from health issues and disease due to poor living conditions and a lack of nearby health facilities. Infrastructure is in poor condition and facilities are generally public or shared (Matamanda and Nel 2021). Informal settlements in India have inadequate waste disposal facilities. The communities rely on communal access to water either as private water taps or shared standpipes near their households (Corburn and Sverdlik 2019). All this negatively impacts community health. Even in the face of disasters, informal settlers have limited ability to cope and bounce back from shocks and stresses as illustrated by Bhanye (2023) showing the impacts of COVID-19 on the poor in Harare, Zimbabwe. In addition to inadequate water and sanitation facilities, there is no access to affordable health facilities (Matamanda et al. 2022).

Informal settlement dwellers suffer a disproportionate burden of ill health. Poor housing, overcrowding, indoor air pollution, lack of sanitation and infrastructure coupled with poverty and vulnerability reinforce poor physical and mental health. This contributes to a double burden of enhanced vulnerability to disease, further entrenching socio-economic and health inequalities (Chirisa et al. 2022). Moving beyond this concern with dwellings' internal health implications, the World Health Organisation's (WHO) holistic approach to housing expands these concerns and identifies four interrelated dimensions, house, home, neighbourhood and community, all contributing to health and well-being.

7.3.2 SDG 13 on Climate Action

The Asia-Pacific region is currently unable to achieve SDG 13 targets. Some vulnerable nations are still trying to recover from the negative impacts of coronavirus and ensure that there is minimum harm to the environment through emissions of GHG (Gero et al. 2017). Climate change and climate-induced disasters fundamentally threaten development in the Asia-Pacific region, undermining the enjoyment of human rights, and worsening societal inequalities by excessively burdening the poor and the most vulnerable groups, including women and children, indigenous populations, gender minorities, migrants, displaced and persons with disabilities. Between 2000 and 2019, eight out of the top ten countries by disaster occurrence and seven out of ten countries by disaster occurrence by disaster-affected population were in Asia-Pacific. The SDG 13 on climate action remains unlikely to be achieved at the current rate of change.

The Intergovernmental Panel on Climate Change (IPCC) cites three factors underpinning vulnerability to climate change exposure to hazards, sensitivity to those hazards, and capacity to adapt. Areas with the greatest exposure and sensitivity to hazards and the least capacity to adapt include large parts of Africa, such as the Sahel, where temperatures are rising more quickly than the global average and a lack of rainfall is already having a profound impact on the environment (Ntema et al. 2021).

South Africa has implemented good policies and strategies to respond to climate change mitigation by ensuring sustainable consumption and production. However, some interventions to mitigate climate change are ineffective because they are voluntary. There is evidence of the absence of concrete actions to implement them, and inconsistencies in the broader policy framework. Climate change adaptation policies and strategies lack adequate institutions to implement and execute them (Weimann and Oni 2019). The institutional capacity seems to be concentrated at the national level and the system is fragile due to dependence on few experts. From what is being experienced in South Africa, it is evident that not enough is being done to support sustainable consumption and production (SDG 12) closely linked to SDG 13. Hence, it is concluded that policies have not improved production methods by the industry and the energy-intensive industry is not complying with the policies. Continued refinement of the climate mitigation policy for industry compliance, promotes renewable energy, energy efficiency and low carbon development trajectory for South Africa (Dintsi et al. 2020).

South Africa has aligned the climate action sustainable goals with its national development agenda, emphasising adaptation policies and institutional framework. This comes against a background where the country has been accused of bias towards mitigation policies triggered by Long-Term Mitigation Scenarios in 2007 that could have quickened mitigation responses to SDG 13. It emerged that South Africa has policies and strategies to respond to climate change adaptation within the context of SDG 13. However, while policies are in place, they have not translated to real change on the ground and therefore have not enable the country to have adequate climate change resilience. Institutions are scattered, with uneven capacity across sectors and different spheres of government, and weakest at the local government level. It also emerged that mitigation was prioritised for a while over adaptation, with a lack of funding and general awareness. Adaptation measures should not be undertaken in isolation, instead, they should be addressed with the context of other programmes such as risk management and sustainable development.

7.4 Case Study of Mangaung Metropolitan Municipality

For the case study, gridded datasets were used to identify spatiotemporal trends of heavy rainfall and warm spell events in the Mangaung Metropolitan Municipality, with a focus on Bloemfontein and the informal settlements in Botshabelo and Thaba Nchu (Fig. 7.1). In 2019 Mangaung Metropolitan Municipality's population rose to 861,651. Informal settlements in Mangaung Metropolitan Municipality is governed by by-laws as promulgated in Provincial Gazette No 60 of 25 October 2013 (Mangaung Metropolitan Municipality By-laws 2013). The objective is the prevention, regulation, monitoring and control of informal settlements and other related matters within the area of jurisdiction of the MMM. Additionally, it prescribes how the Municipality should address and deal with the issues relating to informal settlements. In 2019, Mangaung's population comprised 86% African,

Fig. 7.1 Mangaung Metropolitan Municipality showing the three study areas, Bloemfontein, Botshabelo and Thaba Nchu

11% White and 4% Coloured groups. Using the lower-bound poverty line, 36.6% of the Municipality's live in poverty. The highest inequality coefficient of 0,62 makes Municipality the most unequal in South Africa.

The Mangaung population of 861,651 people comprising 265,560 households, 47% reside in houses, 11.7% reside in informal dwellings (Shacks) and 6.5% in backyard flats. The report further points that, there are 53 informal settlements in the Mangaung area accommodating more than 28,737 households in informal dwellings and backyard shacks.

The CHRIPS and AgERA5 reanalysis datasets were used for this case study. These datasets contained daily rainfall and temperature data for the study period January 1990 to December 2022, respectively. The CHRIPS is a gridded dataset product, available from 1981–present that merges satellite-derived rainfall values with weather station data and has a square grid resolution of 0.05° (~5 km; Funk et al. 2015). The AgERA5 dataset is a statistically downscaled and bias-corrected product available at a 0.1° resolution (Hersbach et al. 2020; C3S 2020). It is derived from the hourly ERA5 reanalysis that is available at a 0.25° resolution and combines numerical model, satellite and observation data (Gelaro et al. 2017; C3S 2020; Hersbach et al. 2020). Single grid cell values corresponding to the location coordinates were extracted to compute the metrics/indices.

As a first step for the data analyses, R-based Climpact Software, developed by World Meteorological Organisation (WMO) Expert Team on Sector-specific Climate Indices (ET-SCI), was used to calculate the heavy rainfall (r10 mm) and the number of days contributing to a warm spell index ($WSDI_n$; Table 7.2). The ET-SCI user-defined warm spell index ($WSDI_n$) was set to three consecutive days where maximum temperatures (TX) were > 90th percentile (e.g., $WSDI_3$), as studies have indicated that three consecutive days above the 90th percentile (or broadly very hot conditions) may result in a significant rise in the mortality rate (Nairn and Fawcett 2014; Van der Walt and Fitchett 2021).

Table 7.2 Selected ET-SCI heavy rainfall and warm spell index

Index	Definition	Units
r10 mm	Number of days when rainfall is at least 10 mm	Days
$WSDI_n$	The annual number of days contributing to events where n or more consecutive days experience maximum temperatures (TX) > 90th percentile	Days

Adapted from Harold and McComb (2022)

Trend analyses for the two indices were performed from 1990–2022. To test whether the heavy rainfall and warm spell indices were non-stationary and evaluate the trend's magnitude, the non-parametric Kendall's tau correlation coefficient (τ) and Sen's Slope were used, respectively (Kendall 1955; Sen 1986; Allen et al. 2015). The significance (p-value) of the trends was conducted at the 5% level (cf. Karki et al. 2019; Yosef et al. 2019).

7.5 Case Study Results

Overall, the Mangaung Metropolitan experienced an increase in three consecutive warm spell days ($WSDI_3$), with rate increases of 0.26 days/year from 1990–2022 (Fig. 7.2a). Similarly, an increased trend was evident for heavy rainfall days (r10 mm) in the Mangaung Metropolitan. Although no increasing or decreasing rate was evident over the study period (Fig. 7.2b).

Bloemfontein and Botshabelo experienced non-significant increasing trends, with rate increases between 0.23 days/year and 0.41 days/year (Fig. 7.3a, b). The greatest increase in three consecutive warm spell days was experienced in Thaba Nchu (Fig. 7.3c).

Non-significant increasing trends are evident for heavy rainfall days (r10 mm) in Bloemfontein and Botshabelo, with rate increases of 0.12 days/year and 0.07 days/year, respectively (Fig. 7.4a, b). No rate increases were evident in Thaba Nchu (Fig. 7.4c).

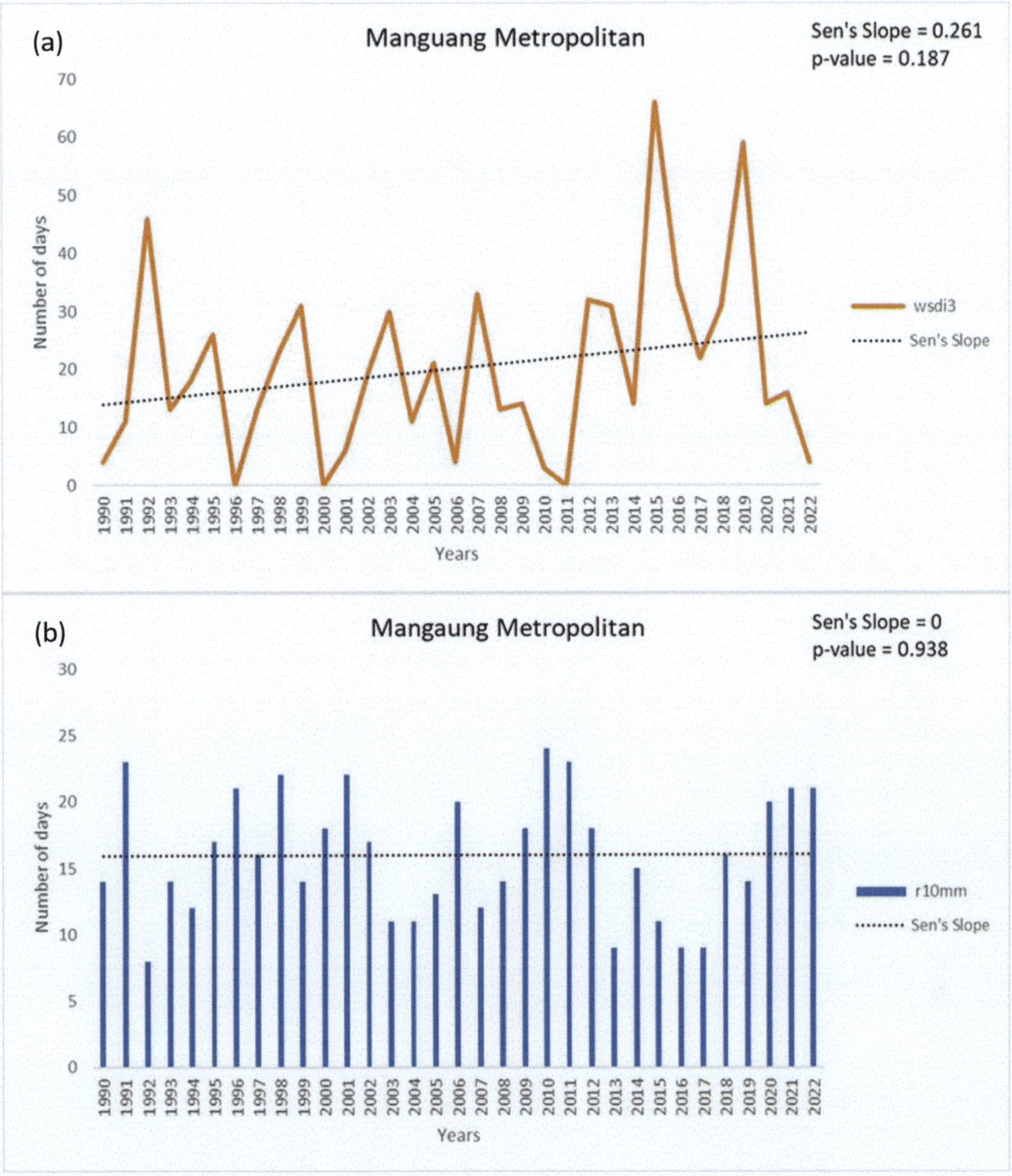

Fig. 7.2 Mangaung Metropolitan (**a**) warm spell days ($WSDI_3$); (**b**) heavy rainfall days (r10 mm)

7.5.1 Impact of Climate Change in Mangaung Metropolitan

The analysis includes extreme climate-related disasters affecting the poor population in the Municipality. This includes the profiling of vulnerable population and low-income groups that reside in environmental risk areas such as along flood plains and in informal settlements (Mokoena and Marais 2007). Like other municipalities Mangaung Municipalities is faced with the challenges of informal settlements, with houses constructed using inferior materials and located in ecologically sensitive areas such as wetlands and flood lines.

Mangaung faces stormwater drainage sanitation problems, especially in highly populated areas with fragile ecosystems. The Mangaung

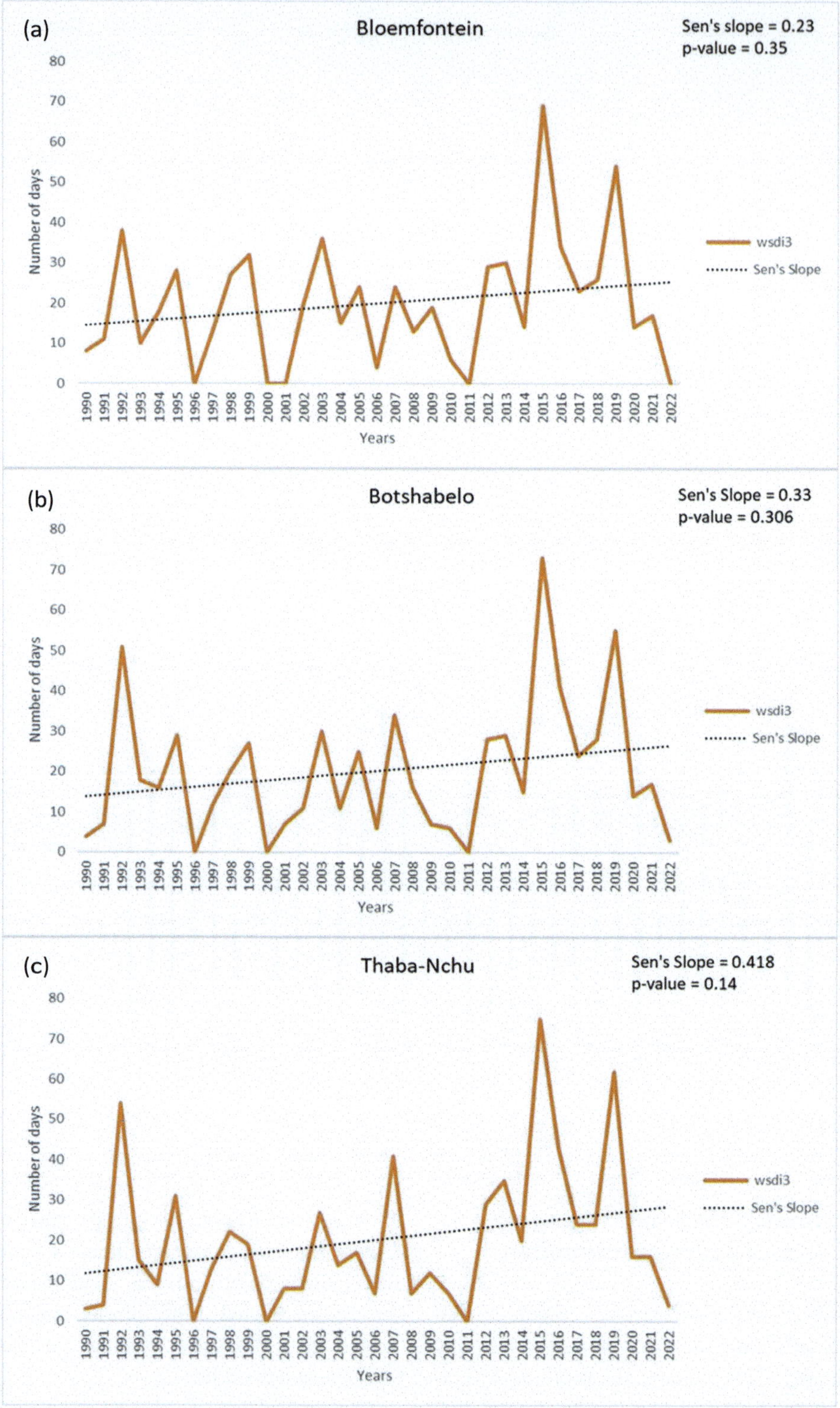

Fig. 7.3 Three consecutive warm spell days ($WSDI_3$) in (**a**) Bloemfontein; (**b**) Botshabelo; (**c**) Thaba Nchu

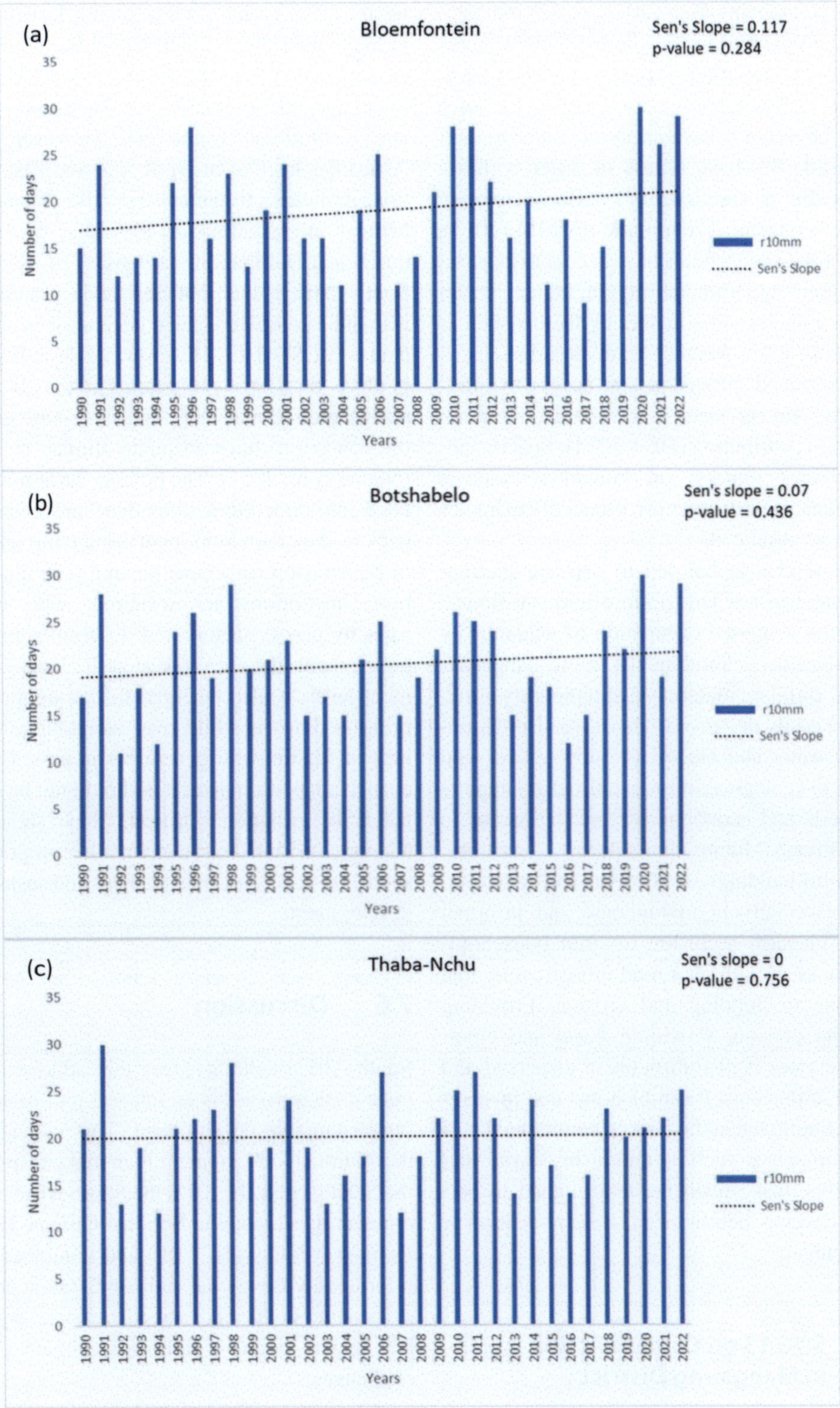

Fig. 7.4 Heavy rainfall days (r10 mm) in (**a**) Bloemfontein; (**b**) Bosthabelo; (**c**) Thaba Nchu

Human Settlements Plan (2013, p. 32) indicates that 40% of main arterial roads and 60% of access streets need upgrading. Progress has been made in providing households with water and as such 92% of households have access to water in their yards while 8% have access to water within a 200 m radius as stipulated in the Reconstruction and Development Programme (MMM 2013). Most (91.4%) households have access to electricity for lighting. Vulnerability of human settlements is influenced by factors by factors such as settlements type, whether rural or urban, peri-urban or mixed, provision and access to infrastructure and services and spatial temporal population distribution (DEA 2014). Specific climate change impacts on human settlements include heatwave discomfort, especially in homes that are not insulated.

Climate change has led to extreme weather conditions that has led to more frequent floods. Due to the increased occupation of wetlands by informal settlers, flooding has more dangerous impacts. Building materials and temporary structures are easily destroyed. There are also disruptions to water and electricity supply. Loss and damage to ecological infrastructure that supports livelihoods and economic sectors that occur to those already living in squalid conditions. Damage to buildings (roof, door, and windows) occurs especially in low-income and informal houses. Physical seclusion of rural households occurs in areas with poor road infrastructure that are prone to flooding and erosion. Providing basic and erosion. Providing basic and emergency services is also difficult in dispersed and isolated settlements. It can be noted that involvement of traditional authorities in communal lands and complexities such as tenure insecurity and land rights, make residents less aligned to personally develop their homes fearing relocation or displacement.

7.5.2 SDG 13 on Climate Action in Mangaung District

The alignment of the SDGs with national development agendas has gained traction since ratifying the 2030 Agenda for Sustainable Development in September 2015. South Africa has aligned the climate action SDG 13 with its national development agenda, emphasising adaptation policies and institutional framework. However, South Africa has been accused of bias towards mitigation policies triggered by the Long-Term Mitigation Scenarios in 2007 that could have quickened mitigation responses to SDG 13. South Africa has policies and strategies to respond to climate change adaptation within the context of SDG 13. However, while policies are in place, they have yet to translate to real change on the ground and therefore, have note enabled the country to have adequate climate resilience (Ntema et al. 2021). The policies have not translated into concrete actions there are knowledge gaps in the adaptation, poor leadership and lack of clear vision for adaptation and poor coordination. Institutions are scattered, with uneven capacity across sectors and different spheres of government that are weakest at the local government level. It also emerged that mitigation was prioritised for a while over adaptation, with a lack of funding and general awareness. Climate change adaptation measures should not be undertaken in isolation. Instead, they should be addressed within the context of other programmes such as disaster risk management and sustainable development.

7.6 Discussion

South Africa acknowledges that addressing climate change issues is an integral part of achieving sustainable development. This is evident in the National Development Plan that encapsulates the SDG targets. In 2011, the South African government developed the National Climate Change Response Policy (NCCRP) and continues to be informed by the global calls on Climate change, such as the IPCC AR5 and AR6. Social equity is key as the country addresses climate change challenges.

According to the National Development Plan, rising temperatures, erratic rainfall and extreme weather events were on the increase and likely to take a heavy toll on Southern Africa. There is expected to be an increased spread of tropical diseases and growing losses (human and financial) from droughts and flooding. Climate change can potentially reduce food production and the availability of potable water that can result in migration patterns and increased conflict.

The Nation, Friday 27 January 2023, reported that South Africa experienced a heat wave with temperatures rising above 40 °C. The heat wave continued for a whole week with eight deaths attributed to heat stroke. Other health challenges included breathlessness, difficulty in sleeping and failure to concentrate due to excessive heat. The heat wave affected the Northern Cape and KwaZulu Natal. Both animals and humans were affected to the extent of collapse. The devastating impact of the extreme climate events were worsened by the power outages where cooling was impossible. In informal settlements that were not fully serviced and equipped, the effects were worse and may have gone unnoticed.

Residents of Ward 8, Grasslands in Bloemfontein are constantly surrounded by pools of sewage. The sewage pools are worsened during flooding leading to air and water pollution and affecting the general ambience of the locations. The unpleasent situation adversely affects the residents through air and water polltion among other problems, thus posing a health risk. Inhaling unpleasant odour of stagnant sewage everyday affects their general daily activities. and may lead to mental health through distress. Blocked drainage pipes and drainage since 2020 have led to this dire state. Coupled with extreme rainfalls, the situation worsens when sewer mixes with drinking water due to old and dilapidated sanitation infrastructure. The changes in extreme rainfall days noted in the three cases studies calls for planning that addresses sewerage disposal and drainage in the study area.

The N8 informal residents protested against poor service delivery which included clean water, house allocation among other needs. This shows how desperate informal settlements residents are in terms of basic sanitary services. During the protest the residents pointed out that they were subjected to long periods of waiting for house allocations which in some cases stretch to more than 15 years.

As a result of economic conditions and unemployment, people are forced to live in such conditions despite the associated risks. Informal settlers, in most cases, have no reliable source of income and health insurance; the people have no money to relocate or afford formal housing (Greibe et al. 2023). The study mentioned how residents of informal settlements experience climate change and its health impacts. Informal settlements are usually located in low-lying areas lacking water and sanitation, health facilities, no stormwater drainage and temporary building materials used in housing construction. These factors and existing environmental degradation and emissions of greenhouse gases, increase the risk of extreme weather conditions, thereby negatively affecting the health of residents.

The residents in informal settlements have a high knowledge of climate change and its health-related impacts. Their attitude towards and motivation to learn about climate change and related health risks to mitigate the effects is present and high. Having lived in the informal settlement and living close to a riverside was significantly associated with having knowledge of climate change, experiencing the impact of climate changes on the community, and being concerned with health-related impacts of climate change. The findings from this study may be used to identify ways in which knowledge on climate change resilience, related to health, can be improved and how education can be tailored to the local context. Further, some insights may prove useful for policy makers, intervention planners and researchers as increased recognition of the impact of climate change on health advocates for local level climate change mitigation and adaption strategies. Thus, the findings may help inform appropriate climate change mitigation and adaptation interventions and local, regional, and national policy making.

Managing change is essential for human survival; thus, the importance of adapting to climate change has been increasingly recognised by researchers and governments alike (Erwin 2009; Melore and Nel 2020). This is reflected in the growing literature on climate change and the imperative for action, including building resilience in socio-ecological systems (Matthews et al. 2015). The study analysed community health factors in the context of climate change regarding those living in precarious housing conditions. In conclusion, people living in informal settlements are at high risk from the negative impacts of climate change. This is because of the location of informal settlements and the absence of adequate service delivery. As the people are low-income earners, they cannot cope in terms of shocks and stresses. This increases their vulnerability. Proactive, collaborative and persuasive action is necessary to create communities without slums.

7.7 Conclusion

The study concludes that attaining SDG 13 means that policy coverage and institutional spread are in place to address climate change adaptation capacity and resilience building. Policies and strategies implemented to curb the effects of climate change should be examined to improve policy practice and even decision-making by the government on adaptation measure. The study revealed that South Africa has policies and strategies in place even though it was inferred that they are superficial, difficult to implement and not supported by resources, some of the policies are failing due to a lack of efficient and effective leadership and a clear vision for adaptation. Climate change adaptation and mitigation policies have not assisted South Africa in building climate change resilience. There was also a concern about poor institutional coordination. Although climate change adaptation is regarded as important in South Africa, there is a general lack of emphasis on adaptation. Hence, adaptation is neglected considering all these challenges and considering South Africa's development agenda, the study recommends that there should be deliberate enhancement of adaptation measures at all levels of government. Furthermore, a full alignment of climate action goals is necessary, specifically the adaptation and resilience perspective should be addressed within the context of other programmes such as disaster risk management and sustainable development.

References

Allen SMJ, Gough WA, Mohsin T (2015) Changes in the frequency of extreme temperature records for Toronto, Ontario, Canada. Theor Appl Climatol 119:481–491

Badland H, Pearce J (2019) Liveable for whom? Prospects of urban liveability to address health inequities. Soc Sci Med 232:94–105

Bhanye J (2023) 'Emerging forms of spatialised and socialised authority' among 'tenure-insecure peri-urbanites' in African peri-urban spaces: a review study. Pan-African Conversations 1(2):26–63

C3S (2020) Copernicus climate change service climate data store. https://doi.org/10.24381/cds.553b7518 (Accessed on 31 Aug 2023)

Chirisa I, Mutambisi T, Chivenge M, Mabaso E, Matamanda AR, Ncube R (2022) The urban penalty of COVID-19 lockdowns across the globe: manifestations and lessons for Anglophone sub-Saharan Africa. GeoJournal 87:815–828

Corburn J, Sverdlik A (2019) Informal settlements and human health. In: Nieuwenhuijsen M, Khreis H (eds) Integrating human health into urban and transport planning: a framework. Springer Nature, Cham, pp 155–171

Department of Environmental Affairs (DEA) (2014) http://www.mangaung.co.za/wpcontent/uploads/2014/07/Environmental-Health-Services-By-Laws.pdf

Dino IG, Akgül CM (2019) Impact of climate change on the existing residential building stock in Turkey: an analysis on energy use, greenhouse gas emissions and occupant comfort. Renew Energy 141:828–846

Dintsi S, Mbanga S, Smallwood J (2020) COVID-19 pandemic and the challenges of informal settlements in South Africa and beyond: theoretical insights and practical interventions. J Public Adm 55(3):283–293

Erwin KL (2009) Wetlands and global climate change: the role of wetland restoration in a changing world. Wetl Ecol Manag 17(1):71–84

Funk C, Peterson P, Landsfeld M, Pedreros D, Verdin J, Shukla S, Michaelsen J (2015) The climate hazards infrared precipitation with stations—a new environmental record for monitoring extremes. Sci Data 2(1):1–21. http://www.mangaung.co.za/wp-content/uploads/2023/06/Annexure-Q-Mangaung-Metro-Informal-Settlement-Upgrading-Strategy-Final-November-2013.pdf

Gajjar SP, Singh C, Deshpande T (2019) Tracing back to move ahead: a review of development pathways that constrain adaptation futures. Clim Dev 11(3):223–237

Gelaro R, McCarty W, Suárez MJ, Todling R, Molod A, Takacs L, Randles CA, Darmenov A et al (2017)

The modern-era retrospective analysis for research and applications, version 2 (MERRA-2). J Clim 30(14):5419–5454
Gero A, Kohlitz J, Willetts JR (2017) Informal settlements in the Pacific and links to sustainable development. Dev Bull 78:91–95
Greibe AJ, Kallestrup P, Karekezi C, Yonga G, Kraef C (2023) Climate change and health risks in Mukuru informal settlement in Nairobi, Kenya–knowledge, attitudes and practices among residents. BMC Public Health 23(1):393
Harold N, McComb J (2022) ARCCSS-extremes: Climpact. GitHub. https://github.com/ARCCSS-extremes/climpact
Hersbach H, Berrisford P, Hirahara S, Horányi A, Muñoz-Sabater J et al (2020) The ERA5 global reanalysis. Q J R Meteorol Soc 147(730):1999–2049
Hoeltl A, Brandtweiner R, Bates R, Berger T (2020) The interactions of sustainable development goals: the case of urban informal settlements in Ethiopia. Int J Sustain Dev Plan 15(3):287–294
Ibe GO, Amikuzuno J (2019) Climate change in sub-Saharan Africa: a menace to agricultural productivity and ecological protection. J Appl Sci Environ Manag 23(2):329–335
Isaacs L (2023) Rivers overflow, people displaced as Cape Town battered by heavy rain. Eyewitness, 15 June. Available from: https://ewn.co.za/2023/06/15/rivers-overflow-people-displaced-as-cape-town-battered-by-heavy-rain (Accessed 1 Oct 2023)
Karki R, Hasson SU, Gerlitz L, Talchabhadel R, Schickhoff U, Scholten T, Böhner J (2019) Rising mean and extreme near-surface air temperature across Nepal. Int J Climatol 40:1–19
Kendall MG (1955) Rank correlation methods. Charles Griffen, London
Lee H, Calvin K, Dasgupta D, Krinner G, Mukherji A, Thorne P, Trisos C, Romero J, Aldunce P, Barrett K, Blanco G (2023) Climate Change 2023: synthesis Report. Contribution of Working Groups I, II and III to the Sixth Assessment Report of the Intergovernmental Panel on Climate Change
Linares C, Díaz J, Negev M, Martínez GS, Debono R, Paz S (2020) Impacts of climate change on the public health of the Mediterranean Basin population-current situation, projections, preparedness and adaptation. Environ Res 182:109107
Madhanagopal D, Jacob VA (2022) Climate risks in an unequal society: the question of climate justice in India. In: Madhanagopal D, Beer CT, Nikku BR, Pelser AJ (eds) Environment, climate, and social justice: perspectives and practices from the global south. Springer Nature, Cham, pp 161–188
Mangaung Human Settlements Plan (2013) Mangaung Metropolitan Integrated Development Plan: Review 2013–2014. Bloemfontein: Mangaung Metropolitan Municipality
Mangaung Metro Municipality (MMM) (2013) http://www.mangaung.co.za/2013/12/
Matamanda AR, Nel V (2021) (un) healthy cities: reflections on urban public health in poor neighbourhoods. In: Matamanda AR, Nel V, Chirisa I (eds) Urban geography in postcolonial Zimbabwe: paradigms and perspectives for sustainable urban planning and governance. Springer Nature, Cham, pp 93–118
Matamanda AR, Dunn M, Nel V (2022) Broken bridges over troubled waters: COVID-19 and the urban poor residing in Dinaweng informal settlement, Bloemfontein, South Africa. S Afr Geogr J 104(3):309–327
Matthews T, Lo AY, Byrne JA (2015) Reconceptualizing green infrastructure for climate change adaptation: barriers to adoption and drivers for uptake by spatial planners. Landsc Urban Plan 138:155–163
Melore TW, Nel V (2020) Resilience of informal settlements to climate change in the mountainous areas of Konso, Ethiopia and QwaQwa, South Africa. Jàmbá: J Dis Risk Stud 12(1):1–9
Mentaschi L, Duveiller G, Zulian G, Corbane C, Pesaresi M, Maes J, Stocchino A, Feyen L (2022) Global long-term mapping of surface temperature shows intensified intra-city urban heat Island extremes. Glob Environ Chang 72:102441
Mokoena M, Marais L (2007) The missing link in cooperative governance and housing delivery: lessons from Mangaung local municipality. Urban Forum 18:311–327
Nairn JR, Fawcett RJ (2014) The excess heat factor: a metric for heatwave intensity and its use in classifying heatwave severity. Int J Environ Res Public Health 12(1):227–253
Ntema J, Anderson I, Marais L (2021) Housing and Possible Health Implications in Upgraded Informal Settlements: evidence from Mangaung Township, South Africa. Housing and SDGs in Urban Africa, pp.71–85
Satterthwaite D, Archer D, Colenbrander S, Dodman D, Hardoy J, Mitlin D, Patel S (2020) Building resilience to climate change in informal settlements. One Earth 2(2):143–156
Sen PK (1986) Estimates the regression coefficient based on Kendall's tau. J Am Stat Assoc 63:1379–1389
Tandon A (2022) Climate change made extreme rains in 2022 South Africa floods 'twice as likely'. CarbonBrief, 13 May. https://www.carbonbrief.org/climate-change-made-extreme-rains-in-2022-south-africa-floods-twice-as-likely/ (Accessed 25 Dec 2023)
Van der Walt AJ, Fitchett JM (2021) Exploring extreme warm temperature trends in South Africa: 1960–2016. Theor Appl Climatol 143(3–4):1341–1360
Van Diemen E (2023) Why more than two months worth of rain fell in Western Cape weekend storm. Daily Maverick, 26 September. Available from: https://www.dailymaverick.co.za/article/2023-09-26-why-more-than-two-months-worth-of-rain-fell-in-western-cape-weekend-storm/ (Accessed 1 Oct 2023)
Weimann A, Oni T (2019) A systematised review of the health impact of urban informal settlements and implications for upgrading interventions in South Africa, a rapidly urbanising middle-income country. Int J Environ Res Public Health 16(19):3608

Yosef Y, Aguilar E, Alpert P (2019) Changes in extreme temperature and precipitation indices: using an innovative daily homogenized database in Israel. Int J Climatol 39:5022–5045

Zhang L, Ameca EI, Cowlishaw G, Pettorelli N, Foden W, Mace GM (2019) Global assessment of primate vulnerability to extreme climatic events. Nat Clim Chang 9(7):554–561

Zhao Q, Yu P, Mahendran R, Huang W, Gao Y, Yang Z, Ye T, Wen B, Wu Y, Li S, Guo Y (2022) Global climate change and human health: pathways and possible solutions. Eco-Environ Health 1(2):53–62

Examining the Impact of Urban Green Space Changes on Public Health

8

Simbarashe Jombo

8.1 Introduction

Ecosystem services from urban green spaces, like parks and residential green areas, significantly enhance public health and well-being (Nesbitt et al. 2017; Stoia et al. 2022). They have been recognised for lowering the incidence of disease and mortality rates among city residents (Villeneuve et al. 2012). By offering psychological relaxation, stress relief, and promoting social interaction (Peters et al. 2010), physical activity (Mullaney et al. 2015), and reducing exposure to harmful pollutants, noise, and excessive heat (Dizdaroglu et al. 2009), urban green spaces have a positive impact on both physical and mental health. The reasons behind these effects are multifaceted, mainly attributed to the diverse ecological and social roles of these spaces (Egorov et al. 2016).

Therefore, including the maintenance of urban green spaces in urban health management is vital for enhancing these health benefits, and supporting sustainable urban development initiatives. The health benefits of urban green spaces are linked to their structural and compositional characteristics, including availability, accessibility, configuration, and vegetation composition (Huang et al. 2017). Health benefits are linked to several Sustainable Development Goals (SDGs). Availability measures whether urban residents can access green spaces, while accessibility considers their proximity to residential areas (Kronenberg et al. 2020). Studies show that a greater presence of green spaces in residential areas is associated with lower mortality rates and lower levels of depression, anxiety and stress among residents (Braubach et al. 2017; Villeneuve et al. 2012).

Accessibility to urban green spaces is commonly defined as the correlation between the distance from residential areas, commercial areas and workplaces to the access points of these spaces (Stoia et al. 2022). This accessibility is of such significance that the World Health Organization (WHO) established a public health indicator to measure it (Annerstedt Van Den Bosch et al. 2016). The WHO recommends that green spaces, at least one ha in size and situated within 300 m of residential areas, should be the accessibility benchmark (Huang et al. 2017). Urban green space accessibility is vital as it fosters increased biodiversity, serves as a noise barrier, and provides a nature-like environment, which is essential for physical and mental health (Iraegui et al. 2020). Remote sensing methods have been used to assess the accessibility and availability of urban green spaces using benchmark values (Huang et al. 2017; Iraegui et al. 2020; Malika et al. 2021). This underscores the importance of examining urban green space

S. Jombo (✉)
Department of Physical and Earth Sciences, School of Natural and Applied Sciences, Sol Plaatje University, Kimberley, South Africa

A. R. Matamanda, V. Nel (eds.), *Sustainable Development Goals and Urban Health*, Sustainable Development Goals Series, https://doi.org/10.1007/978-3-031-68734-1_8

accessibility and availability in the pursuit of achieving the SDGs.

Urban green spaces contribute significantly to multiple SDGs, including 15, 3, 11, and 13. The accessibility and availability of urban green spaces align with SDG 15 by providing essential urban ecosystems and supporting biodiversity. These spaces also aid in preserving biodiversity, acting as noise barriers, and offering nature-like environments (Iraegui et al. 2020). Furthermore, urban green spaces have the potential to advance SDG 13, focusing on climate action, as they absorb carbon dioxide, mitigate urban heat effects, and enhance climate resilience (Mullaney et al. 2015; Nesbitt et al. 2017; Wolch et al. 2014). The accessibility of urban green spaces, crucial for physical and mental well-being, is closely linked to the pursuit of good health and well-being for all, as emphasised in SDG 3 (Wey et al. 2022). These spaces are integral to achieving SDG 11, promoting sustainable and inclusive cities. Policymakers and urban planners should prioritise their development and preservation in the broader sustainable development context.

Remote sensing provides an efficient method to monitor urban green spaces, yielding data on their extent and changes (Byomkesh et al. 2012). Previous studies have employed remote sensing to analyse urban green space accessibility and availability. For instance, Su et al. (2019) used satellite imagery to investigate the impact of different green space calculations on urban health in Barcelona, Spain, finding a significant association with improved health outcomes. Kabisch and Haase (2014) studied urban green space accessibility and availability in Berlin, Germany. The results indicated that while many regions had an above-average allocation of urban green spaces compared to the target of 6 m^2 per person, there were notable disparities based on age and immigrant status. Huang et al. (2017) evaluated the shift in health benefits provided by urban green spaces in 28 global megacities from 2005 to 2015, using proxy indicators of accessibility and availability. The study results indicated that the availability of urban green spaces in megacities rose from 27.63% in 2005 to 31.74% in 2015. Similarly, accessibility to urban green spaces increased from 65.76% in 2005 to 72.86% in 2015. This increase in availability and accessibility contributed to increased health benefits for residents of these megacities.

South Africa demonstrated early support for the 2030 Agenda for sustainable development, with a strong alignment between the country's National Development Plan (NDP) and the SDGs. However, there is room for improvement, as South Africa currently ranks 110th out of 166 countries, with an SDG Index score of 64 out of 100 (Sachs et al. 2023). The SDGs encompass various dimensions, including social, economic, environmental, governance, peace, justice, and security. The social SDG 3 focuses on health and well-being, while the environmental SDGs, such as SDGs 11, 13, and 15, address sustainable cities, climate action, and conservation of biodiversity, respectively. Although many developed countries have made progress towards achieving SDGs 3, 11, 13, and 15, South Africa still faces significant challenges in these areas (Salvia et al. 2019). Notably, SDGs 3 and 15 present ongoing obstacles, while SDGs 11 and 13 also require concerted efforts for meaningful progress (Sachs et al. 2023). South Africa's challenges highlight the need to address issues related to urban green space availability and access, ensuring that all residents can enjoy the benefits of these spaces. By prioritising the development, maintenance and equitable distribution of urban green spaces, South Africa can make significant strides in the advancement and creation of sustainable, inclusive, and healthy cities for its population.

Despite numerous studies on urban green space characteristics, their potential health benefits have been largely understudied. The qualities of urban green spaces have been linked to health benefits and can serve as rough indicators to assess their health-related advantages. The WHO recommended standards for access to urban green space, which highlight their potential benefits. Remote sensing technology plays a crucial role in evaluating the attributes and changes of urban green spaces and, by extension, their potential health benefits, due to their broad coverage and data availability. However, research in this area has been limited. The city of

Bloemfontein in South Africa has lost vegetation and urban spaces over the years due to urbanization, changing land use, population growth, infrastructure development, and natural disasters like drought (Dingaan and Du Preez 2017; Sinxadi and Campbell 2020; Wu et al. 2023). This highlights the imperative of enhancing urban green space access and development in the city to improve the health and well-being of its residents. Ensuring equitable access to urban green spaces aligns with significant goals for sustainable and healthy development.

In this chapter, availability and accessibility were used to investigate the change in health benefits offered by urban green spaces in Bloemfontein, South Africa, between 2010 and 2020. The specific objectives of the study were to i) quantify the availability of urban green spaces in Bloemfontein in 2010 and 2020, using Landsat images, ii) to determine the accessibility of urban green spaces in Bloemfontein in 2010 and 2020, using extracted urban green spaces and population grid data, and iii) and discuss the potential health implications of changes in the availability and accessibility of urban green spaces in the study area. The findings are mainly related to SDG 3, 13 and 15 (Fig. 8.1).

Fig. 8.1 Link between sustainable development goals (SDGs) and the availability and accessibility of urban green spaces

Subsequently, the paper presents the materials and methods section, which encompasses a detailed description of the study area, data sources, and image classification techniques employed. Additionally, the validation data source and the materials used in the study are highlighted. Moving forward, the research findings are presented, encompassing the generated Land Use and Land Cover (LULC) maps, as well as an analysis of the availability and accessibility of urban green spaces. These findings serve as the basis for future discussions, in which the significance, limitations, and final remarks of the study are deliberated, offering valuable insights for further research and application in the field.

8.2 Materials and Methods

8.2.1 Bibliometric Analysis

A comprehensive bibliographic analysis was performed using data extracted from the Web of Science (WOS) database, chosen for its vast scale, including more than 2.2 billion cited references, more than 34,000 peer-reviewed journals, and 254 subject categories (Clarivate 2023). The search query encompassed these relevant terms: "ecosystem services," "urban health", "city health", "metropolitan health", "urban well-being," "urban population health", "urban public health", "urban green space" and "ecosystem services." The initial search yielded 226 publications, including conference proceedings, editorial materials, and data papers. After filtering the documents, a total of 201 articles were selected and used for analysis in this study. These articles were published between 1995 and 2023 (Fig. 8.2). There was a significant increase in the number of publications per year, from 1 in 1995 to 12 in 2023, focusing on studies related to ecosystem services, urban health, and urban green spaces (Fig. 8.2). As the search concluded at the end of June 2023, it is expected that the number of publications ($n = 12$) and the average times cited per year ($n = 15$) for 2023 will likely increase by the end of the year (Fig. 8.2). The more frequently a word appeared in the text, the larger the word size

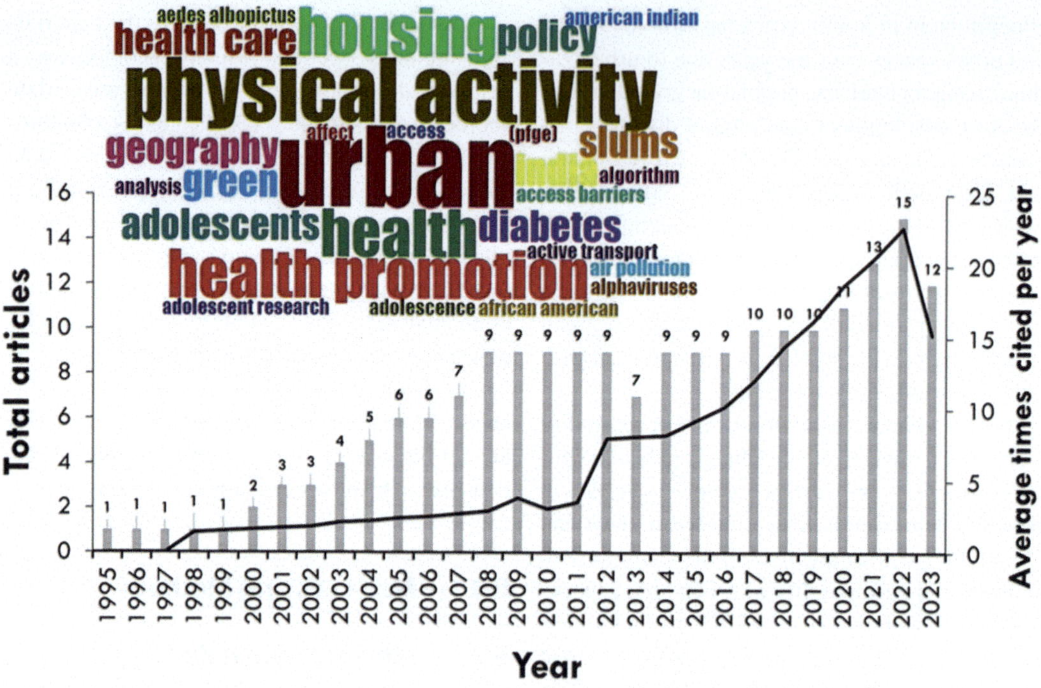

Fig. 8.2 Total number of articles, citations per year and word cloud from 1995 to 2023

in the word cloud. Among the frequently used keywords in these publications, "urban" was the most prevalent, followed by "physical activity" (Fig. 8.2). The findings indicate that studies have increased over the years and human health is greatly being considered in urban areas.

8.2.2 Study Area

The study was conducted in the city of Bloemfontein, located in the Free State province with the city centre located at 29.1208 S and 26.2376 E (Fig. 8.3). The city covers 236.2 km^2 with an altitude between 1350 and 1450 m above sea level. The annual mean maximum and minimum temperatures are 24.6 °C and 7.6 °C, respectively (South African Weather Service 2023). The area typically receives rain during the summer in the form of thunderstorms, with an average of 550 mm per year. In 2023, the population of the Bloemfontein metropolitan area increased by 1.7% from 2022, reaching a total of 598,000 residents (Macrotrends 2023).

8.2.3 Data Collection and Methods

This study used secondary data to evaluate the distribution of urban green spaces and estimate their accessibility. The distribution of urban green space was determined by analysing satellite images using supervised and post-classification methods. To calculate accessibility, the proxy distance between urban green spaces and built-up areas was measured. A comprehensive review of the scientific literature, interviews, and discussions was conducted to determine the most appropriate proxy distance for assessing the accessibility of urban green spaces. Previous studies for example, Nor et al. (2017) and Khan et al. (2023) also employed this technique to assess accessibility.

Samples for six LULC types, namely bare land, impervious surfaces, vegetation, and water bodies, were obtained from Landsat imageries. High-resolution images from Google Earth were used as references during the collection process. A total of 822 samples were used to classify 2010 and 2020 imageries (Table 8.1). Samples from

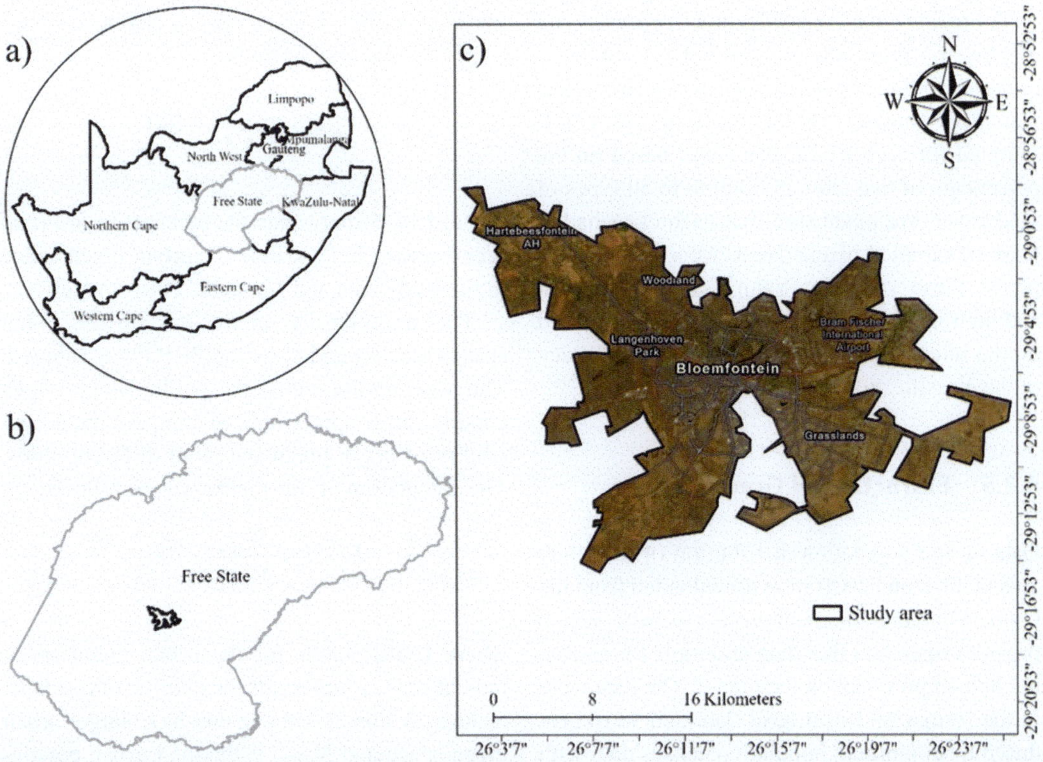

Fig. 8.3 Study area locality at (**a**) national, (**b**) provincial and (**c**) local contexts

Table 8.1 Total number of training and test samples for Land Use and Land Cover (LULC) classes in 2010 and 2020

	Year	2010		2020	
Class		Training samples	Test samples	Training samples	Test samples
Bare land		70	30	81	35
Impervious surface		82	35	92	39
Vegetation		100	43	122	52
Waterbody		66	28	73	31
Sum		**318**	**136**	**368**	**157**

both years were randomly divided into two groups, 70% were used for training and 30% for testing, as shown in Table 8.1. Other studies (e.g., Jombo et al. 2020b; Jombo et al. 2022; Odindi et al. 2016) observed that using a 70/30 data split led to the highest level of classification accuracy and the lowest standard deviation.

8.2.4 Satellite Imagery Classification and Validation

The random forest (RF) algorithm on the Google Earth Engine (GEE) platform was used to classify the Landsat satellite imageries for 2010 and 2020. Breiman (2001) developed the RF classification method, which combines multiple tree predictors. In this approach, each tree relies on a value from a vector that is randomly selected and evenly distributed throughout the forest (Jombo et al. 2020a). The algorithm can process a large amount of training data and variables and still maintain accuracy even when some of the data are missing (Breiman 2001). In this study, 100 trees were used, and the number of variables used in each split was set to the square root of the total number of variables.

Following the classification process, we employed a proportional stratified random sampling technique to extract 100 pixels from each classified image. The proportion of pixels sampled for each LULC class was based on the percentage of that class in relation to all types of land cover. We made sure that each class had at least 10 sampled pixels. Next, we visually interpreted these validation samples using Landsat and high-resolution Google Earth images. Based on the interpretation, we estimated the accuracy of the classification results.

8.2.5 Extraction of Green Spaces

Data on the population distribution of the study area in 2010 and 2020 were downloaded from the Global Human Settlement project website, which provides open and free data and tools for assessing human presence on the planet. The data were in the format of 100-m grid data and were produced by combining population census data with built-up grid data. The urban built-up extent boundary for each year was delineated in the study area using the LULC maps. Initially, intensity maps were produced for different types of LULC, including bare land, impervious surface, vegetation, and water bodies. These maps were generated by moving-window analysis in ArcGIS Pro 3.3 (Esri Inc., Redlands, CA, USA), with a window size of 1 km by 1 km. We applied a multi-criteria approach to ascertain if a pixel, with a resolution of 1 km, was part of the built-up class. Subsequently, we extracted vegetation cover within the boundary of the built-up area from the LULC map. We then integrated this information with the existing vegetation class in the study area to delineate urban green spaces.

8.2.6 Computation of the Availability and Accessibility of Green Spaces in Urban Areas

In this study, the Percentage of Urban Green Spaces (PUGS) was calculated as a measure of the availability of green spaces. To do this, we used the following formula:

$$PUGS = \frac{UGS_i}{UA_i} \times 100$$

where PUGS represents the percentage of urban green space, UGS_i stands for urban green space in the study area and UA_i is the built-up area.

The accessibility metric for urban green spaces recommended by the WHO was adopted. The accessibility metric has public health implications and is appropriate to compare the accessibility of urban green spaces in different cities. The calculation of this metric was as follows:

$$UGSI = \frac{N_{ACC}}{N_{TOTAL}} \times 100$$

where UGSI stands for the urban green space indicator, N_{ACC} represents the number of residents residing within 300 m of a one-ha or larger urban green space and N_{TOTAL} is the total urban population within the confines of the built-up area. To estimate the accessibility of urban green spaces in the study area, all patches of such spaces with a minimum area of one ha were identified. The patches buffered by 300 m were patched and the number of people residing within the buffered areas was computed to obtain N_{ACC}. The total urban population was obtained by counting the number of individuals residing within the boundary of built-up areas, which was denoted as N_{TOTAL}.

8.2.7 Statistical Approaches

The paired t-test was used to determine the significance of changes in availability and accessibility indicators for urban green spaces between 2010 and 2020. This study considered two climatic variables (annual precipitation and annual mean temperature) and two socioeconomic variables (population density and gross domestic product (GDP) per capita). Pearson's correlation analysis was performed to investigate the relationship between explanatory factors (climatic and socioeconomic factors) and the availability and accessibility of urban green spaces. Next, we conducted a multiple general linear model regression analysis to investigate how climatic and socioeconomic factors influenced the availability and accessibility of urban green spaces.

8.3 Results

8.3.1 LULC Mapping and Validation

The LULC maps indicated the area with the most vegetation was mainly in the central section of the study area in both 2010 and 2020 (Fig. 8.4). There was an increase in land covered by impervious surfaces over the 10 years of study, however, there was a decline in vegetation and waterbodies (Fig. 8.4). The overall accuracy values were 87% and 91% for 2010 and 2020, respectively, which exceeded the suggested threshold of 85% (Anderson 1976). The values of the kappa coefficient were 0.86 and 0.90 for 2010 and 2020, respectively.

8.3.2 Urban Green Space Availability

The presence of urban green spaces was observed throughout the built-up boundary in both 2010 and 2020, as shown in Fig. 8.5. However, there was a noticeable decrease in the percentage of urban green spaces by 2020 (Fig. 8.5).

The PUGS experienced a decrease of 2.09%, declining from 21.43% in 2010 to 19.43% in 2020. Similarly, the UGSI exhibited a decline of 6.72%, dropping from 54.35% in 2010 to 47.63% in 2020.

The multiple general linear model regression for availability demonstrated a strong connection between the presence of urban green spaces and climate variables in 2010 and 2020, as shown in Table 8.2. Additionally, the analysis did not reveal a significant correlation between green space availability and socioeconomic factors during the study period. In particular, the p-values for climate variables were all below 0.05, whereas the p-values for the socioeconomic variables exceeded 0.05. There was a negative correlation between availability and population density in 2010 (−0.888), indicating that as population

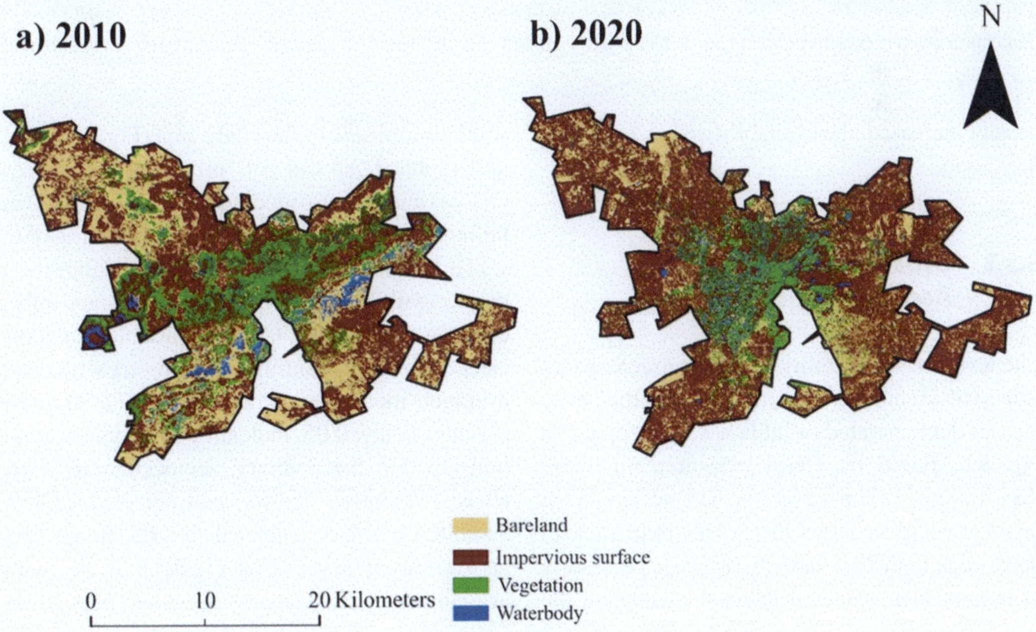

Fig. 8.4 Land Use and Land Cover (LULC) maps for (**a**) 2010, and (**b**) 2020

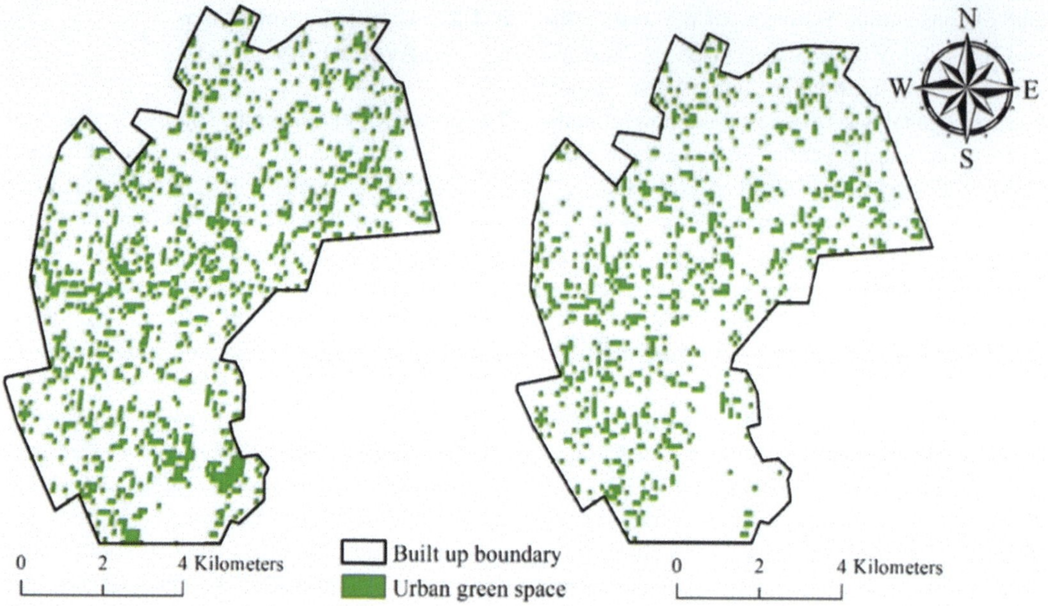

Fig. 8.5 Distribution of urban green spaces in 2010 and 2020

Table 8.2 Findings from multiple regression analysis using a general linear model for availability in both 2010 and 2020. The annual mean rainfall (AMR) and the annual mean temperature (AMT) are the climate variables

2010				2020		
Variable	Coefficient	t	p-value	Coefficient	t	p-value
Intercept	22.041	1.829	0.110	118.599	3.660	0.008
AMR	0.003	0.037	0.012	−0.299	−3.983	0.005
AMT	1.676	1.169	0.021	0.723	1.227	0.030
Population density	−0.888	−4.969	0.802	−0.001	−3.222	0.215
Per capita gross domestic product	0.154	0.348	0.738	−0.428	−0.919	0.389

density increased, the availability of green spaces decreased.

8.3.3 Urban Green Space Accessibility

The level of accessibility to urban green spaces exhibited significant variation across the study area, as demonstrated in Table 8.3 and depicted in Fig. 8.6. Based on UGSI measurement, more than 10% of urban residents had access to an urban green space larger than 1 ha within a 300 m distance in both 2010 and 2020. The accessibility of urban green spaces decreased mainly on the southeastern side of the study area (Fig. 8.6). The geographical spread of urban green spaces (≥1 ha) and the surrounding areas within a 300 m buffer zone in the study area is shown in Fig. 8.6.

The climate variables had a notable impact on the accessibility of green spaces, like their influence on availability. Climate variables, specifically average monthly rainfall (AMR) and average mean temperature (AMT) showed p-values below 0.05, indicating a significant relationship. On the contrary, socioeconomic variables, population density, and per capita GDP, exhibited p-values greater than 0.05, suggesting no significant association (Table 8.3). As with availability in 2010, accessibility and population

Table 8.3 Findings of multiple regression analysis using a general linear model for accessibility in 2010 and 2020. The annual mean rainfall (AMR) and the annual mean temperature (AMT) are the climate variables

	2010			2020		
Variable	Coefficient	t	p-value	Coefficient	t	p-value
Intercept	139.001	3.002	0.015	86.024	3.581	0.001
AMR	−10.072	−2.901	0.040	−10.092	−1.706	0.013
AMT	11.870	2.546	0.006	12.348	−3.794	0.045
Population density	−0.879	0.182	0.148	−0.795	−0.691	0.504
Per capita gross domestic product	−0.160	−0.326	0.753	0.453	1.308	0.231

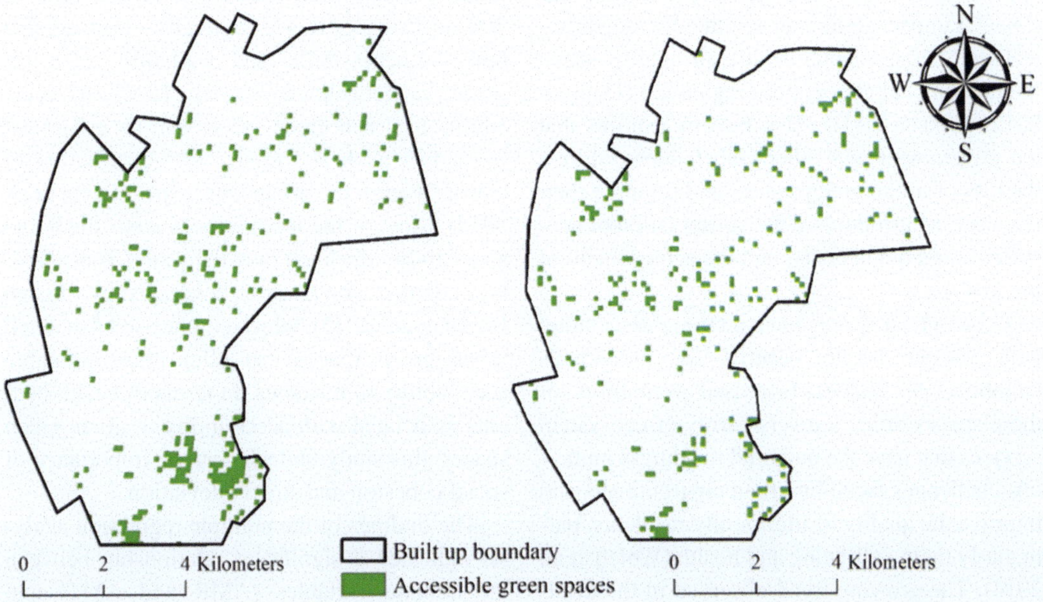

Fig. 8.6 Spatial arrangement of urban green spaces (> 1 ha) and surrounding areas within a 300 m buffer in 2010 and 2020

density showed a strong negative correlation in both 2010 (−0.879) and 2020 (−0.795).

8.4 Discussion

8.4.1 Availability of Urban Green Spaces and the Impact on Health Benefits

The availability of urban green spaces exhibited variation between 2010 and 2020, influenced by natural and socioeconomic factors. The results of the multiple regression analysis indicated a significant inverse relationship between higher average mean temperature (AMT) and the availability of urban green spaces in the study area (Table 8.2 and Fig. 8.5). In contrast, higher mean annual rainfall levels (AMR) exhibited significantly higher availability of urban green spaces (Table 8.2 and Fig. 8.5). This is in line with Liu et al. (2023) who highlighted that high precipitation and sustained temperatures inhibit the growth of vegetation. There was a significant negative correlation between the availability of green spaces and population density. This is consistent with McDonald et al. (2023) who found a negative correlation between population density and the availability of urban green spaces. Moreover, the study revealed that there was no significant correlation between the availability of urban green spaces and per capita GDP. This finding is consistent with the research conducted by Astell-Burt et al. (2014), who emphasised that areas with a higher proportion of low-income residents tended to have limited access to green spaces.

The availability of urban green spaces decreased by 2.09% between 2010 and 2020. This is supported by Senbore and Oke (2023) who found that urban green spaces were converted into impervious surfaces in the city of Bloemfontein. The loss of green spaces has a detrimental impact on human health, affecting physical well-being, mental health, air quality, social interaction, and temperature regulation (Douglas 2012; Jackson 2003). This pattern requires that the government and other stakeholders address the loss of urban green spaces and promote their recovery, contributing to the conservation of ecosystems and biodiversity, and the achievement of the SDGs.

Urban green spaces have a strong association with various health benefits for individuals (Jombo et al. 2020a). Increased availability of these spaces offers a diverse array of ecosystem services that have the potential to address numerous challenges faced by urban environments and improve the quality of life for city residents, particularly their well-being and health (Wolch et al. 2014). The growing trend observed in the availability of urban green spaces suggested an increasing contribution to the overall well-being of residents (Nesbitt et al. 2017). However, the city of Bloemfontein faces a challenge to reap health benefits from urban green spaces due to the limited availability. The pressing issues of population increase, urban growth, severe air pollution, and the urban heat island problem further underscore the crucial need for the health-promoting effects of urban green spaces in Bloemfontein (Jombo and Adelabu 2022). Addressing urban green space availability, accessibility and their associated health consequences has been a long-standing priority in international development efforts (Wei 2018). Incorporating urban greening initiatives into comprehensive urban development strategies is crucial for Bloemfontein to work effectively toward achieving SDG 15, which focuses on sustainable land use and the protection of ecosystems.

8.4.2 Accessibility of Urban Green Spaces and the Impact on Health Benefits

Based on our analysis using the UGSI, we observed a decline in the accessibility of urban green spaces by 6.72% between 2010 and 2020 (Table 8.2). In both 2010 and 2020, more than 10% of the residents had convenient access to urban green spaces larger than 1 ha within a 300 m distance. It was generally observed that both the accessibility and availability of urban green spaces declined over time. However, accessibility to urban green spaces can be influenced by factors such as the size and distribution of patches within the urban landscape (Huang et al. 2017). Most of the urban spaces were small and accessibility declined over the years. These findings contrast with research conducted by van den Bosch et al. (2015), which emphasised that small urban green spaces can offer more equitable accessibility to residents. In terms of health benefits associated with accessibility to urban green spaces, this study underscores the importance of social cohesion and stress alleviation.

The findings of the multiple regression analysis indicated a significant relationship between the climate variables (AMR and AMT) and accessibility to urban spaces (Table 8.3). These findings are in agreement with Bille et al. (2023) who highlighted that climate variables (rainfall and temperature) have a relative influence on the accessibility of urban green space, with a positive effect of precipitation and a negative effect of high temperatures. In contrast, there was no significant relationship between socioeconomic variables and accessibility to urban green spaces. This is supported by the findings of Sathyakumar et al. (2019) who indicated that there was a nonlinear relationship between the quality and accessibility of green spaces with the uniform increase or decrease with socioeconomic status. These results show that urban areas where annual precipitation values are high have more accessible urban green spaces than the socioeconomic status of a place would predict.

In general, the availability of urban green spaces decreased between 2010 and 2020, despite not meeting the accessibility standard set by the WHO. A significant majority, comprising more than 80% of the population, was unable to meet the WHO accessibility standard during both periods. The combination of rapid urbanization, vegetation loss, and population growth (Sinxadi and Campbell 2020) is expected to exacerbate the challenge of limited accessibility to urban spaces. Enhancing accessibility to urban green spaces is therefore crucial, aligning to promote sustainable development (SDG 15) and addressing the urgent need for improved accessibility to support the well-being of the community.

The level of accessibility to urban green spaces decreased between 2010 and 2020 in which 10% of urban residents had access. This decrease might have been due to urbanization and population growth. This is supported by Huang et al. (2022) who highlighted that urban green space accessibility declined with a changing LULC due to urbanization and a subsequent decline in green spaces. Ensuring that urban green spaces offer health benefits is dependent on their accessibility (Annerstedt Van Den Bosch et al. 2016; Huang et al. 2017; Stoia et al. 2022). Greater accessibility in cities empowers urban residents to enjoy these advantages, emphasising the need to prioritise accessibility in green space planning and development (Stoia et al. 2022). This underscores the importance of enhancing accessibility in urban areas, like Bloemfontein, through measures such as improved transportation infrastructure, pedestrian-friendly pathways, and strategic green space placement (Fan et al. 2022).

Enhanced accessibility aligns with SDGs 3, 13, and 15 and supports SDG 3 by promoting residents' physical and mental well-being through green space access (Geary et al. 2021). It addresses SDG 13 by reducing the urban heat island effect, curbing carbon emissions, and bolstering climate resilience (Lorenzo-Sáez et al. 2021). Lastly, it contributes to SDG 15 by preserving biodiversity, enhancing ecosystem services, and promoting sustainable land use practices (Lorenzo-Sáez et al. 2021; Wey et al. 2022). In summary, prioritising and enhancing availability and accessibility in urban green space planning, including in Bloemfontein, is vital for residents to fully enjoy the related health benefits and to advance multiple SDGs, resulting in improved health, climate resilience, and sustainable land use practices.

8.4.3 Limitations of the Study

In assessing the change in health benefits derived from urban green spaces in Bloemfontein, we combined availability and accessibility indicators. Population data were obtained from the Global Human Settlement project that reports every five years, with the latest data available for 2020. Although our goal was to use more recent data, we had to rely on 2020 data for both LULC classification, as well as urban green space availability and accessibility. Remote sensing proves to be a reliable method for accurately capturing and quantifying spatiotemporal changes in the availability and accessibility of urban green spaces. However, the use of medium-resolution remote sensing data, such as Landsat images, poses challenges in effectively capturing human activities occurring within urban green spaces and their associated health benefits. Additionally, medium or coarse-resolution remote sensing data fails to capture the presence of small-scale urban green spaces, which play a significant role in providing health benefits. Although medium-resolution remote sensing datasets offer valuable insights at a larger scale, the use of high-resolution remote sensing data would be beneficial, particularly in areas like Bloemfontein, to obtain more precise measurements of green spaces for increased accuracy in analyses.

8.5 Conclusion

Urban green spaces provide several health benefits to urban residents. Their availability and accessibility with a growing urban population are essential for residents. Gaining insight into the scope and spatial distribution of health benefits is essential to strategically plan and design urban green spaces, intending to maximise the positive impact on people's well-being. This study introduces a novel approach that uses remote sensing data to rapidly assess the health benefits offered by urban green spaces in the city of Bloemfontein, South Africa. Two indicators were used to quantify the availability and accessibility of these green spaces. The analysis involved extensive use of Landsat images and the RF classifier, enabling precise assessments of the characteristics and changes in urban green spaces. The findings revealed a decline in both the availability and accessibility of urban green spaces from 2010 to 2020. This suggests that most city residents do not acquire as many benefits as could be provided by urban green spaces. Moving forward, the priority of the creation of new green spaces and the preservation of existing ones should be emphasised to ensure a sustained positive impact on public health and well-being. This study highlights the critical role of urban green spaces in improving SDGs 3, 13, and 15, which, respectively, focus on ensuring good health and well-being, combating climate change, and protecting and restoring ecosystems on land. The availability and accessibility of urban green spaces are essential to address the needs of urban residents, particularly with the ever-growing urban population.

References

Anderson JR (1976) A land use and land cover classification system for use with remote sensor data, vol 964. US Government Printing Office

Annerstedt Van Den Bosch M, Mudu P, Uscila V, Barrdahl M, Kulinkina A, Staatsen B, Swart W, Kruize H, Zurlyte I, Egorov AI (2016) Development of an urban green space indicator and the public health rationale. Scand J Public Health 44(2):159–167

Astell-Burt T, Feng X, Mavoa S, Badland HM, Giles-Corti B (2014) Do low-income neighbourhoods have the least green space? A cross-sectional study of Australia's most populous cities. BMC Public Health 14(1):292

Bille RA, Jensen KE, Buitenwerf R (2023) Global patterns in urban green space are strongly linked to human development and population density. Urban For Urban Green 86:127980

Braubach M, Egorov A, Mudu P, Wolf T, Ward Thompson C, Martuzzi M (2017) Effects of urban green space on environmental health, equity and resilience. In: Nature-based solutions to climate change adaptation in urban areas: linkages between science, policy and practice. SpringerOpen, pp 187–205

Breiman L (2001) Random forests. Mach Learn 45(1):5–32

Byomkesh T, Nakagoshi N, Dewan AM (2012) Urbanization and green space dynamics in Greater Dhaka, Bangladesh. Landsc Ecol Eng 8:45–58

Clarivate (2023) Web of science core collection. https://clarivate.com/webofsciencegroup/solutions/web-of-science-core-collection/ (Accessed 21 July 2023)

Dingaan MN, Du Preez PJ (2017) Floristic composition and species diversity of urban vegetation in Bloemfontein, Free State, South Africa. Bothalia-African Biodivers Conserv 47(1):1–15

Dizdaroglu D, Yigitcanlar T, Dawes LA (2009) Sustainable urban futures: an ecological approach to sustainable urban development. In: In proceedings of the second infrastructure theme postgraduate conference 2009: rethinking sustainable development-planning, infrastructure engineering, design and managing urban infrastructure, 2009. Queensland University of Technology, pp 187–195

Douglas I (2012) Urban ecology and urban ecosystems: understanding the links to human health and Well-being. Curr Opin Environ Sustain 4(4):385–392

Egorov AI, Mudu P, Braubach M, Martuzzi M (2016) Urban green spaces and health-a review of evidence. WHO Regional Office for Europe, Copenhagen

Fan PY, Chun KP, Mijic A, Tan ML, Liu MS, Yetemen O (2022) A framework to evaluate the accessibility, visibility, and intelligibility of green-blue spaces (GBSs) related to pedestrian movement. Urban For Urban Green 69:127494

Geary RS, Wheeler B, Lovell R, Jepson R, Hunter R, Rodgers S (2021) A call to action: improving urban green spaces to reduce health inequalities exacerbated by COVID-19. Prev Med 145:106425

Huang C, Yang J, Lu H, Huang H, Yu L (2017) Green spaces as an indicator of urban health: evaluating its changes in 28 mega-cities. Remote Sens 9(12):1266

Huang Y, Lin T, Zhang G, Jones L, Xue X, Ye H, Liu Y (2022) Spatiotemporal patterns and inequity of urban green space accessibility and its relationship with urban spatial expansion in China during rapid urbanization period. Sci Total Environ 809:151123

Iraegui E, Augusto G, Cabral P (2020) Assessing equity in the accessibility to urban green spaces according to different functional levels. ISPRS Int J Geo Inf 9(5):308

Jackson LE (2003) The relationship of urban design to human health and condition. Landsc Urban Plan 64(4):191–200

Jombo S, Adelabu S (2022) Spatiotemporal variations of land surface temperature and vegetation coverage in Free State Province, South Africa. In: IGARSS 2022–2022 IEEE international geoscience and remote sensing symposium, 2022. IEEE, pp 2414–2417

Jombo S, Adam E, Byrne MJ, Ali KA, Newete SW (2020a) Exploring the potential of feature selection methods in the classification of urban trees using field spectroscopy data. Int J Geospat Environ Res 7(3):3

Jombo S, Adam E, Byrne MJ, Newete SW (2020b) Evaluating the capability of Worldview-2 imagery for mapping alien tree species in a heterogeneous urban environment. Cogent Soc Sci 6(1):1754146

Jombo S, Adam E, Tesfamichael S (2022) Classification of urban tree species using LiDAR data and WorldView-2 satellite imagery in a heterogeneous environment. Geocarto Int 37(3):1–28

Kabisch N, Haase D (2014) Green justice or just green? Provision of urban green spaces in Berlin, Germany. Landsc Urban Plan 122:129–139

Khan MW, Hossain MS, Alam M (2023) GIS-based analysis to identify the distribution and accessibility of urban green space in Dhaka Metropolitan City, Bangladesh. J Geogr Inf Syst 15(1):35–52

Kronenberg J, Haase A, Łaszkiewicz E, Antal A, Baravikova A, Biernacka M, Dushkova D, Filčak R, Haase D, Ignatieva M, Khmara Y, Niţă MR, Onose DA (2020) Environmental justice in the context of urban green space availability, accessibility, and attractiveness in postsocialist cities. Cities 106:102862

Liu L, Peng J, Li G, Guan J, Han W, Ju X, Zheng J (2023) Effects of drought and climate factors on vegetation dynamics in Central Asia from 1982 to 2020. J Environ Manag 328:116997

Lorenzo-Sáez E, Lerma-Arce V, Coll-Aliaga E, Oliver-Villanueva J-V (2021) Contribution of green urban areas to the achievement of SDGs. Case study in Valencia (Spain). Ecol Indic 131:108246

Macrotrends (2023) Bloemfontein, South Africa Metro Area Population 1950–2023. https://www.macrotrends.net/cities/22479/bloemfontein/population (Accessed 03 Mar 2023)

Malika O, Boudjemaa K, Salim D, Abla KA (2021) Quantitative and qualitative assessment of urban green spaces in Boussaada City, Algeria using remote sensing techniques. J Geogr Reg Plann 14(3):123–133

McDonald RI, Aronson MF, Beatley T, Beller E, Bazo M, Grossinger R, Jessup K, Mansur AV, Puppim de Oliveira JA, Panlasigui S (2023) Denser and greener cities: Green interventions to achieve both urban density and nature. People Nat 5(1):84–102

Mullaney J, Lucke T, Trueman SJ (2015) A review of benefits and challenges in growing street trees in paved urban environments. Landsc Urban Plan 134:157–166

Nesbitt L, Hotte N, Barron S, Cowan J, Sheppard SRJ (2017) The social and economic value of cultural ecosystem services provided by urban forests in North America: a review and suggestions for future research. Urban For Urban Green 25:103–111

Nor ANM, Corstanje R, Harris JA, Brewer T (2017) Impact of rapid urban expansion on green space structure. Ecol Indic 81:274–284

Odindi J, Mutanga O, Rouget M, Hlanguza N (2016) Mapping alien and indigenous vegetation in the KwaZulu-Natal Sandstone Sourveld using remotely sensed data. Bothalia 46(2)

Peters K, Elands B, Buijs A (2010) Social interactions in urban parks: stimulating social cohesion? Urban For Urban Green 9(2):93–100

Sachs J, Kroll C, Lafortune G, Fuller G, Woelm F (2023) Sustainable development report 2023. Cambridge University Press

Salvia AL, Leal Filho W, Brandli LL, Griebeler JS (2019) Assessing research trends related to Sustainable Development Goals: local and global issues. J Clean Prod 208:841–849

Sathyakumar V, Ramsankaran R, Bardhan R (2019) Linking remotely sensed Urban Green Space (UGS) distribution patterns and Socio-Economic Status (SES)—a multi-scale probabilistic analysis based in Mumbai, India. GISci Remote Sens 56(5):645–669

Senbore S, Oke SA (2023) Urban development impact on climate variability and surface water quality in part of Mangaung metropolis of South Africa. Dev South Afr 40(2):293–312

Sinxadi L, Campbell M (2020) Factors influencing urban open space encroachment: the case of Bloemfontein, South Africa. In: Smart and sustainable cities and buildings. Springer Nature, p 285

South African Weather Service (2023) Bloemfontein. https://www.weathersa.co.za/ (Accessed 03 Mar 2023)

Stoia NL, Niţă MR, Popa AM, Iojă IC (2022) The green walk—an analysis for evaluating the accessibility of urban green spaces. Urban For Urban Green 75:127685

Su JG, Dadvand P, Nieuwenhuijsen MJ, Bartoll X, Jerrett M (2019) Associations of green space metrics with health and behavior outcomes at different buffer sizes and remote sensing sensor resolutions. Environ Int 126:162–170

van den Bosch M, Mudu P, Uscila V, Barrdahl M, Kulinkina A, Staatsen B, Swart W, Kruize H, Zurlyte I, Egorov A (2015) Development of an urban green space indicator and the public health rationale. Scand J Public Health 44

Villeneuve PJ, Jerrett M, Su JG, Burnett RT, Chen H, Wheeler AJ, Goldberg MS (2012) A cohort study relating urban green space with mortality in Ontario, Canada. Environ Res 115:51–58

Wei Y (2018) Urban expansion, sprawl and inequality. Landsc Urban Plan 177:259–265

Wey YE, Sarma V, Lechner AM, Nath TK (2022) Malaysians' perception on the contribution of urban green spaces to the UN sustainable development goals. Urban For Urban Green 78:127792

Wolch JR, Byrne J, Newell JP (2014) Urban green space, public health, and environmental justice: the challenge of making cities 'just green enough'. Landsc Urban Plan 125:234–244

Wu S, Chen B, Webster C, Xu B, Gong P (2023) Improved equality of human exposure to greenspace in the 21st century urbanization. Nat Commun 14(1)

9 Peace, Justice and Strong Institutions: Reflections on Health Governance and Health Inequalities in Mangaung

Abraham R. Matamanda, Fortune Mangara, Tafadzwa Maramura, and Charles P. Mukumba

9.1 Introduction

The Sustainable Development Goal (SDG) # 16 envisages promoting peaceful and inclusive societies for sustainable development, providing access to justice for all, and building effective, accountable and inclusive institutions at all levels by 2030. This SDG addresses a critical issue relating to institutional dynamics, social inclusiveness, and how it impacts urban health. The significance of institutions is highlighted by the role of different stakeholders holding power, influence and authority that can influence decision-making and ultimately, resource allocation and access.

Considering the complexity of cities, socio-spatial exclusion and marginalisation tend to be prevalent, resulting in the neglect of specific social groups that fail to access certain goods and services in cities. For example, residents in informal settlements are usually excluded from accessing essential services such as water, sanitation and public transport, which eventually impact health outcomes (Matamanda et al. 2022). On the other hand, politics of difference may result in favouritism or prioritisation of some social groups while marginalising others.

This chapter focuses on SDG 16 and provides the South African context of this SDG, recognising the strides and challenges related to achieving the set targets. The link to urban health is examined at the national level, followed by a local-level analysis that articulates the status quo in Mangaung Metropolitan Municipality (MMM) and how SDG 16 impacts urban health in the metro. The chapter contributes to understanding the institutions' role in spurring the SDGs' success. This is critical because the achievement of the SDGs considers the institutions' role in spurring good governance, transparency, accountability, and urban liveability. A qualitative research approach was adopted in developing this chapter, and data was collected from secondary sources. Specifically, data was collected from municipal reports, policy documents, spatial development frameworks for MMM and newspapers focusing on SDG# 16 and urban health in the context of MMM.

A. R. Matamanda (✉)
Department of Urban & Regional Planning, University of the Free State, Bloemfontein, South Africa

F. Mangara
Department of Urban & Regional Planning, North West University, Potchefstroom, South Africa

T. Maramura
Department of Public Administration, University of the Free State, Bloemfontein, South Africa

C. P. Mukumba
Center for Development Support, University of the Free State, Bloemfontein, South Africa

A. R. Matamanda, V. Nel (eds.), *Sustainable Development Goals and Urban Health*, Sustainable Development Goals Series, https://doi.org/10.1007/978-3-031-68734-1_9

9.2 The Conceptual Framework of SDG 16

The conceptual framework of SDG 16 in this context recognises the interconnectedness of peace, justice, and strong institutions in achieving sustainable development and the varied nuances. It further acknowledges that peace and security, access to justice, and effective governance are essential for eradicating poverty, promoting economic growth, and achieving other sustainable development goals. First, peace and security promote peaceful and inclusive societies, prevent violence, and reduce conflict. Amin and Graham (1997) assert that peace and security include vital aspects such as reducing violence and related death rates, ensuring access to justice, strengthening the rule of law, and combating organised crime and corruption.

Access to justice is the second pillar, highlighting the importance of equal access to justice for all individuals and communities. Harvey (1973) confirms that access to justice emphasises the need for responsive, inclusive, transparent legal systems and effective and accountable institutions at all levels. Therefore, access to justice entails key objectives such as increasing legal identity for all, providing legal aid and support to vulnerable groups, and promoting the rule of law (Lima and Gomez 2021; Sommerlad 2004).

Third, effective, accountable, and inclusive institutions help advance good governance at all levels. This pillar also emphasises the need for transparency, accountability, and inclusiveness in decision-making processes (Omri and Mabrouk 2020). The key elements which underpin this pillar include promoting participatory governance, reducing corruption, ensuring public access to information, and enhancing capacity-building for effective institutions (Denters et al. 2023; United Nations 2015).

Overall, the conceptual framework of SDG 16 serves as a roadmap for creating peaceful, just, and inclusive societies with accountable institutions that uphold the rule of law and protect human rights. Furthermore, the conceptual framework acknowledges that progress in SDG 16 is crucial for the overall success of the 2030 Agenda. It also recognises the importance of partnerships and collaboration among governments, civil society organisations, and other stakeholders to implement effective strategies and policies to achieve SDG 16 targets, as illustrated below through the SDG Index and Dashboards Report (Fig. 9.1).

Despite SDG 16 primarily focusing on promoting peaceful, just, and inclusive societies, some goals, targets, and indicators relate to urban health within this broader framework. While urban health is not the central focus of SDG 16, it also intersects with several aspects of the goal, as tabulated below through extraction from the SDG Index and Dashboards Report (SDG Index and Dashboards Report 2022).

Table 9.1 shows that institutional dynamics influence health outcomes in an area. While these elements within SDG 16 touch upon urban health, it is worth noting that there are other SDGs, such as SDG 3 (Good Health and Well-being), SDG 6 (Access to clean water and sanitation) and SDG 11 (Sustainable Cities and Communities) (United Nations 2015), that also explicitly address health and urban issues more directly. These goals provide a more comprehensive framework for understanding and addressing urban health challenges.

9.3 Theoretical Underpinnings: The Socio-Spatial Justice Theory

Graziano and Ruggiero (2023) and Soja (2010) highlight that the socio-spatial justice theory is an approach that explores the intersections between social justice and spatial or geographical dimensions. The theory examines how the distribution of resources, opportunities, and social outcomes is shaped by spatial factors and how spatial inequalities can perpetuate social injustices (Kühn 2014). So, the socio-spatial justice theory and SDG 16, which focuses on promoting peaceful, just, and inclusive societies, share common objectives and concerns. Therefore, the socio-spatial justice theory can be applied to Sustainable Development Goal 16 in several ways. SDG 16

Fig. 9.1 Summary of the SDG 16 and the set targets. (Source: SDG Index and Dashboards Report (2022))

aims to promote peaceful and inclusive societies, provide access to justice for all, and build effective, accountable, and inclusive institutions at all levels (United Nations 2009). While SDG 16 addresses broader social and governance issues, the socio-spatial justice theory offers a lens through which to understand and analyse the spatial dimensions of social justice and how they relate to SDG 16.

Socio-spatial justice theory recognises the role of social movements and advocacy in addressing spatial injustices (Sclar et al. 2005). Similarly, SDG 16 encourages the active participation of civil society, grassroots organisations, and marginalised groups in decision-making processes. Hence, both emphasise the importance of mobilising around spatial issues and advocating for social justice and inclusivity. Socio-spatial justice theory also acknowledges the intersections between spatial inequalities and other forms of social discrimination (Soja 2010). SDG 16 also recognises the importance of access to justice for all individuals, regardless of their social identities. Thus, both underpinnings emphasise the need to address multiple intersecting dimensions of inequality to achieve social justice and inclusive societies. This is emphasised by considering the spatial dimensions of social justice. Socio-spatial justice theory can provide insights into the challenges and strategies for achieving the goals of SDG 16. Therefore, the socio-spatial justice theory helps to identify spatial disparities, understand their underlying causes, and propose interventions that promote more equitable and inclusive societies (Kühn 2014), thus aligning with the aims of SDG 16.

Socio-spatial justice theory emphasises the importance of ensuring equal access to justice for all individuals and communities. Graziano and Ruggiero (2023) affirm that the theory recognises that marginalised groups, such as people with low incomes, ethnic minorities, and people living in remote or disadvantaged areas, may face barriers to accessing justice. Therefore, achieving SDG 16 requires addressing these inequities and ensuring that justice systems are accessible, affordable, and responsive to the needs of all people, regardless of their social or spatial location.

Socio-spatial justice theory highlights the need to fairly distribute institutional resources and capacities across different regions and communities. In the context of SDG 16, this means that effective, accountable, and inclusive institutions should not be concentrated in certain urban areas or wealthy regions while neglecting rural or marginalised communities (Ruijsbroek et al. 2017). Thus, achieving socio-spatial justice requires equitable allocation of resources to build and strengthen institutions at all levels, ensuring that they serve the needs of diverse populations.

Table 9.1 SDG #16 targets and link to urban health

Target and indicators	Urban health connection
Target **16.1:** Significantly reduce all forms of violence and related death rates everywhere.	Urban areas may experience higher levels of violence, including interpersonal violence and crime, which can have a significant impact on public health. Addressing violence in urban contexts is crucial for improving urban health outcomes.
Target 16.3: Promote the rule of law at the national and international levels and ensure equal access to justice for all.	Equal access to justice is important for addressing urban health disparities and ensuring that all residents, including marginalised populations in urban areas, have access to essential healthcare services and legal protections related to health.
Target 16.7: Ensure responsive, inclusive, participatory, and representative decision-making at all levels.	Inclusive and participatory decision-making processes are essential for addressing urban health challenges and promoting health equity in urban areas. Engaging communities and stakeholders in urban planning and policymaking can lead to healthier and more sustainable urban environments.
Indicator 16.3.2: Unsentenced detainees as a proportion of the overall prison population.	The indicator indirectly relates to urban health as the conditions within urban prisons can have implications for the health and Well-being of unsentenced detainees. Overcrowding, inadequate healthcare services, and poor sanitation in urban prisons can contribute to health risks and the spread of infectious diseases.

Socio-spatial justice theory is also important since it recognises that spatial planning and urban development policies can profoundly impact social equity (Graziano and Ruggiero 2023). In the context of SDG 16, it calls for inclusive urban governance and planning processes that consider the needs and aspirations of all residents, particularly marginalised groups. This includes ensuring access to basic services, affordable housing, public spaces, and infrastructure in both urban and rural areas. Thus, the socio-spatial justice theory also emphasises avoiding spatial inequalities and promoting inclusive and sustainable urban development.

The socio-spatial justice theory also calls for inclusive and participatory governance structures that enable all individuals, regardless of their social or spatial location, to have a voice in shaping policies and institutions that affect their lives (Soja 2010). This is because empowering marginalised groups and promoting their active participation can help address power imbalances and ensure their perspectives and interests are considered in justice-related reforms (Baum et al. 2016). In summary, applying socio-spatial justice theory to SDG 16 involves addressing inequities in access to justice, distributing institutional resources fairly, promoting inclusive spatial planning and urban development, and empowering marginalised communities through participation. By incorporating these principles, efforts towards achieving SDG 16 can be more effective in promoting peaceful, just, and inclusive societies (WHO 2018). Therefore, addressing spatial inequalities is crucial for building peaceful societies as it helps to mitigate social tensions, reduce disparities, and promote social cohesion.

9.4 Socio-Spatial Justice Theory, SDG 16 and Urban Health

Socio-spatial justice theory, SDG 16, and urban health are interconnected in several ways. Socio-spatial justice theory recognises the importance of addressing spatial inequalities and injustices in urban contexts (Soja 2010), while SDG 16 aims to promote inclusive, just, and peaceful societies (United Nations 2015). Urban health focuses on improving individuals' and communities' health and well-being (Friel et al. 2011). Hence, the socio-spatial justice theory emphasises the role of spatial inequalities in shaping health outcomes. In urban areas, there are often disparities in access to healthcare services, environmental quality, and social determinants of health, such as housing, education, and employment (Lee and Maheswaran 2011). These disparities make addressing spatial inequalities essential for promoting equitable urban health.

Achieving socio-spatial justice involves ensuring equitable access to healthcare and addressing health disparities in urban settings. This implies that spatial inequalities can manifest in various ways, such as unequal access to resources and services, segregation, gentrification, and spatial disparities in different socioeconomic conditions (Swyngedouw 2004). In many urban areas, marginalised populations face barriers to accessing quality healthcare, including physical and financial barriers, discriminatory practices, and unequal distribution of healthcare facilities (De Villiers 2021; Schmid 2012). This is why SDG 16 calls for equal access to justice and healthcare services. Ultimately, the socio-spatial justice theory examines the causes and consequences of these inequalities and seeks to address them through governance, policy, and action. By incorporating socio-spatial justice principles into the implementation of SDG 16 in urban contexts, it is possible to address health inequalities and promote urban health.

9.5 Context of SDG 16 in SA

It is widely recognised that governments worldwide face significant challenges in achieving all the SDGs simultaneously. Therefore, many have prioritised specific goals and targets aligning with national development plans and objectives. By prioritising certain SDGs, governments can focus their efforts and resources more effectively to achieve the most significant impact and progress towards sustainable development. This approach acknowledges the interconnected nature of the SDGs and recognises that progress towards one goal can have positive spillover effects on others. Ultimately, prioritising specific SDGs enables governments to work towards creating a more sustainable and equitable future while also addressing the unique challenges and circumstances they face at the national level.

SDG 16 is a priority for the South African government as it aligns with the objectives of the National Development Plan (NDP), which aims to promote security, access to justice, and the rule of law while fighting corruption (Statistics South Africa 2019). The NDP's vision for a united, prosperous, non-racial, non-sexist, and democratic country is also closely aligned with the targets of SDG 16. Therefore, the South African government recognises the importance of SDG 16 in achieving its national development goals and creating a society that values equity, justice, and human rights.

According to the South African Sustainable Development Goals Hub (n.d.), SDG 16 is critical to the overall sustainable development agenda. The South African government has acknowledged its importance by prioritising specific targets and indicators to achieve this goal. The South African government has set the following targets to be achieved by 2030:

1. *Strengthen institutions to promote peace, justice, and accountability.*
2. *Reduce all forms of violence and related death rates.*
3. *Ensure equal access to justice for all.*
4. *Develop effective, accountable, and transparent institutions at all levels.*
5. *Substantially reduce corruption and bribery in all their forms.*
6. *Promote the rule of law at the national and international levels and ensure equal access to justice for all.*
7. *Ensure responsive, inclusive, participatory, and representative decision-making at all levels.*
8. *Broaden and strengthen developing countries' participation in global governance institutions.*

By prioritising these targets, the South African government aims to foster an inclusive society while creating an enabling environment for sustainable development.

The South African government has established a set of indicators to measure progress towards achieving the prioritised targets of SDG 16. These indicators cover a range of issues, including public safety, access to legal identity, gender equity in government, and corruption.

Some indicators include the percentage of people who feel safe walking alone in their neighbourhoods at night, the rate of intentional homicide, and the proportion of the population with access to a legal identity (UN 2017). Other indicators focus on gender equity, such as the proportion of female government officials in senior positions. Additionally, the government measures the total amount of illicit financial flows out of the country and the proportion of the population who have experienced at least one bribe payment request in the last 12 months (OECD 2017). These indicators provide a framework for monitoring progress towards SDG 16 and can help guide policy decisions and interventions. By regularly tracking and reporting on these indicators, the government can monitor progress towards its sustainable development objectives and make informed decisions about policies and interventions to drive further progress (IFPRI 2019).

The South African government has taken significant steps towards achieving SDG 16 by implementing various policies and initiatives such as the National Crime Prevention Strategy, National Policy Framework for Children, and National Anti-Corruption Strategy. These policies and initiatives aim to strengthen institutions, promote accountability and transparency, reduce corruption and bribery, and ensure equal access to justice for all. Also, the government has established the South African SDG Hub (n.d) to track progress towards achieving the SDGs, including SDG 16. The Hub serves as a platform to provide information on the country's policies, initiatives, and progress towards achieving the SDGs. It also serves as a resource for collaboration among stakeholders, including government agencies, civil society organisations, private sector partners, and academic institutions.

The South African government's recognition of SDG 16 in achieving sustainable development is reflected in its prioritisation of specific targets and indicators to be achieved by 2030. Implementing policies and initiatives and establishing the South African SDG Hub (n.d) provides a strong foundation for tracking progress towards achieving these targets and collaborating among stakeholders to achieve them.

9.5.1 Progress and Strides Made in SA Towards SDG 16

The South Africa SDG Index and Dashboard Report 2022 (SDG Index and Dashboards – Global Data Platform 2022) provides a comprehensive analysis of South Africa's progress towards achieving the SDGs, highlighting areas where more action is needed to achieve the goals. Figure 9.2 shows the SDG 16 performance by indicator.

According to the SDG Index and Dashboards Report for 2022, South Africa has made progress in some areas related to SDG 16. For example, the homicide rate has decreased from 35.8 per 100,000 people in 2015 to 33.5 in 2020. Additionally, the percentage of children involved in child labour decreased from 5.5% in 2016 to 3.6% in 2019. These improvements suggest that efforts to promote peace and justice and build strong institutions in South Africa have some positive impacts. Despite some progress in some areas related to SDG 16, there are also indicators where South Africa has not shown improvement. For example, the Corruption Perception Index remained unchanged at 44 in 2021 compared to the previous year.

These indicators suggest that corruption and the functioning of the justice system remain significant challenges in South Africa's efforts to promote peace, justice, and strong institutions. There are also indicators related to SDG 16, where South Africa has shown little change or stagnation. For example, the proportion of the population who feel safe walking alone at night in the city or area where they live remained unchanged at 32% since 2017. Property rights have also not improved significantly, with the score remaining 4.3 out of a best possible score of 7 in 2020.

South Africa has made some strides in achieving SDG 16 despite significant challenges, such as high levels of violent crime and corruption. The country has implemented measures to promote peace, justice, and accountability, including strengthening institutions such as the judiciary and the anti-corruption watchdog and improving the criminal justice system. South Africa has also

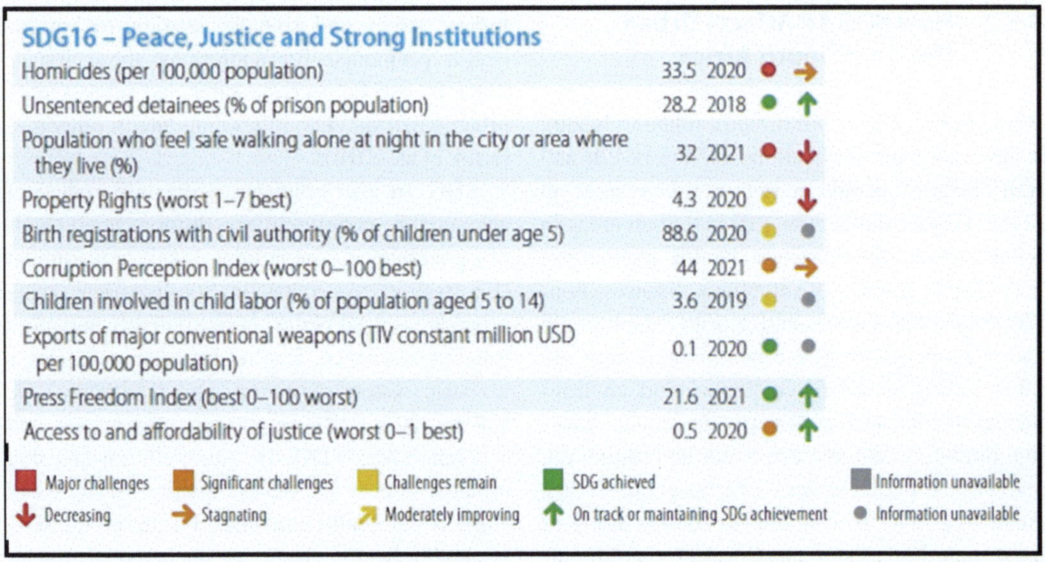

Fig. 9.2 SDG 16. (Source: *SDG Index and Dashboards Report* (2022))

improved access to justice through community-based justice services and legal aid programs. The country also has a relatively high level of press freedom, supported by a robust legal framework and active civil society.

However, SDG 16 intersects with other issues, particularly health governance and health inequalities. The COVID-19 pandemic has highlighted the need for more significant investment in healthcare infrastructure and resources, particularly in marginalised communities, and the importance of addressing social determinants of health that contribute to health inequalities (UNSTATS 2020). Poverty, unemployment, and inadequate housing are some of the underlying factors contributing to poor health outcomes, and addressing these factors is essential to promoting health equity.

South Africa has struggled with health governance issues, including corruption, lack of accountability, and inadequate infrastructure. Addressing these issues will require a comprehensive and coordinated approach from the government, civil society, and the private sector (Government of South Africa 2020). It will also require transparency, accountability, and adherence to sound governance principles. According to the Government of South Africa (2019a, 2019b), the country has addressed these challenges such as implementing a national health insurance scheme to improve access to healthcare services for all citizens and investing in healthcare infrastructure and resources in marginalised communities. The country has also implemented social welfare programs to address the social determinants of health and promote health equity.

To achieve SDG 16, it is essential to address the challenges related to health governance and health inequalities in South Africa (Government of South Africa 2019a, 2019b). The COVID-19 pandemic has highlighted the importance of addressing these issues, and the country has taken steps to promote health governance and address health inequalities. However, these challenges are complex and multifaceted, requiring a sustained effort from all stakeholders involved. Achieving SDG 16 in South Africa will create a healthier and more equitable society, with peace, justice, and strong institutions for all.

Given South Africa's challenges in achieving SDG 16 and the importance of addressing the social determinants of health, it is essential to consider how SDG 16 affects urban health in the country, particularly given the high levels of urbanisation and the concentration of poverty and inequality in urban areas.

9.5.2 How SDG 16 Affects Urban Health in South Africa

While SDG 16 does not directly relate to health, it has important implications for the health and well-being of people living in urban areas in South Africa (Shulla et al. 2021). Urban areas in South Africa face various health challenges, including high rates of infectious diseases, non-communicable diseases, and injuries (Maphumulo and Bhengu 2019). These health challenges are often linked to social, economic, and environmental factors, such as poverty, inadequate housing, limited access to clean water and sanitation, and unsafe public spaces (Zerbo et al. 2020). Achieving SDG 16 can help to address some of these underlying factors and improve health outcomes in urban areas in South Africa (StatsSA 2019). For example, building effective and inclusive institutions can ensure that resources are allocated fairly and transparently and that policies and programs are responsive to the needs of all communities, including those living in urban areas.

Similarly, promoting peaceful and inclusive societies can help reduce violence and crime in urban areas, significantly impacting health outcomes (Cerdá et al. 2018). For example, violence can lead to injuries, mental health problems, and increased stress and anxiety, all of which can have long-term health consequences. In addition, access to justice for all is an essential component of SDG 16. It can help to ensure that people in urban areas have access to the resources and support they need to address health-related issues, such as environmental pollution or inadequate housing.

SDG 16 plays a crucial role in shaping urban health outcomes in South Africa. By promoting peaceful, just, and inclusive societies, SDG 16 can create an enabling environment for improving health equity and reducing health inequalities in urban areas. One of the key targets of SDG 16 is to reduce violence and insecurity in all forms, including physical and sexual violence, as well as political and social violence. In urban areas, violence and crime are major public health concerns that cause physical harm and generate psychological stress and trauma, leading to various health problems. By addressing these issues, SDG 16 can help create safer and more secure urban environments, improving health outcomes (Kruk et al. 2018).

SDG 16 also emphasises the importance of accountable and effective institutions at all levels of governance, including local government (Eilertsen et al. 2020). In South Africa, many urban areas need better service delivery and more robust governance, leading to inadequate access to basic water, sanitation, and healthcare (De Villiers 2021). By improving governance and accountability, SDG 16 can help ensure that urban residents can access quality services and enjoy better health outcomes. Furthermore, SDG 16 highlights the need for inclusive decision-making processes and equal access to justice and legal systems (Chapman and de Langen 2021). In urban areas, marginalised and vulnerable communities often face barriers to accessing justice and participating in decision-making processes, which can exacerbate health inequalities (National Academy of Sciences 2017). By promoting inclusive and participatory approaches, SDG 16 can help empower communities and ensure their health needs are considered (WHO 2017).

SDG 16 is critical for improving urban health outcomes in South Africa by addressing violence and insecurity, promoting accountable governance and service delivery, and ensuring inclusive decision-making processes and access to justice (Government of South Africa 2013).

9.6 Results from Mangaung

9.6.1 Institutional and Governance System in MMM

Mangaung Local Municipality was declared a Metropolitan area in 2011 after successfully hosting the FIFA 2010 World Cup. MMM is a category A metropolitan municipality that is supposed to have autonomy over its finances. The Executive Mayor holds the highest office, followed by the City Manager. A strategic support team complements the mayor and city manager, which includes one general manager and a secretary. As shown in Fig. 9.3, the political offices are positioned parallel to the Executive Mayor and City Manager, showing the undefined roles of politicians. The undefined roles of politicians have negatively affected governance in MMM as political interference in the daily operations of the MMM was prevalent and responsible for misgovernance and corruption. Ten directors report to the City Manager on different issues, including finance, human settlements, corporate services, police services, waste and fleet management.

9.6.2 Protests and Shutdowns

Protests have been rife in MMM over the past years. These protests have largely been due to a lack of information, service delivery, unemployment issues, and frustrations from residents (Nkuyane 2022; Sifile 2021). In May 2021, the Mangaung Concerned Community organised a shutdown protest triggered by the dissatisfaction the residents experienced with the ANC-led Free State government (Mthethwa 2021). The protests were characterised by violence as the protestors in Bloemfontein barricaded and closed major roads, preventing residents from travelling to the city for work and other businesses. As a result, government buildings and schools were closed for two days, and some protestors looted shops operated by foreign nationals whom some accused of taking jobs and economic opportunities from them.

First, the protests have been associated with some deaths, which represents some health challenges. Amid the violence associated with the May 2021 protests, a 15-year-old teenager was shot and killed, depicting the challenges associated with the protests that can result in the death of innocent citizens (Chabalala 2021). Several residents have also lamented that the protests cause some anxiety and emotional trauma as they fear for their lives. This is evident from the violent nature of the protests, which causes some residents to be victimised, especially some foreign nationals who are at times identified as taking opportunities from the locals. The persisting violent protests associated with violence compromise the achievement of SDG#16, which envisages reducing all forms of violence and deaths.

Second, the protests have been associated with complete shutdowns of economic activities in Bloemfontein, which have seen roads being blocked and barricaded and certain activities coming to a halt. In this way, there have been multiple disruptions and inconveniences for residents who fail to access health services. For example, in 2021, the protest in Mangaung resulted in the COVID-19 vaccination programme being halted in some parts of

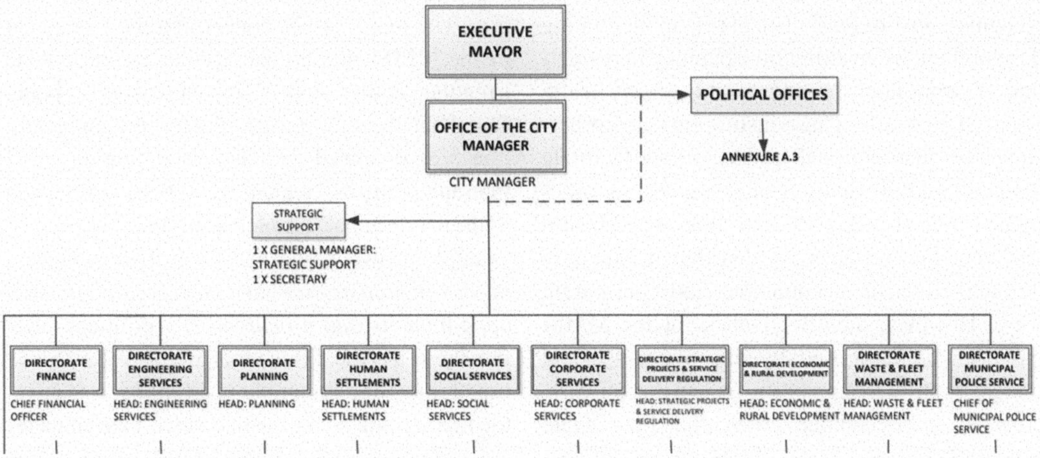

Fig. 9.3 Organisational structure for Mangaung Metro

Bloemfontein that were heavily affected by the violent protests. In this case, Mochoari (2021) reported how the service delivery protest in Mangaung resulted in the COVID-19 vaccination programme being put on hold.

In 2023, the National Education Health and Allied Workers Union (Nehawu) allegedly orchestrated a strike in MMM, calling for salary increases among health workers (GroundUp 2023). While protesting is every citizen's right, some protestors' actions resulted in barricading major referral hospitals in Bloemfontein, making it difficult for healthcare workers to access their workplaces and engage in their professional duties. While the Democratic Alliance (DA) lodged a human rights violation complaint, such actions significantly compromised urban health as patients failed to access healthcare services, resulting in some challenges for sick and ailing citizens (Clarke 2023). These persisting protests in MMM highlight the challenges in meeting the SDG #11 target, which focuses on strengthening institutions to promote peace, justice, and accountability.

9.6.3 Develop Effective, Accountable, and Transparent Institutions at All Levels

The development of effective, accountable and transparent institutions at all levels is echoed among the targets set for SDG#16. This target recognises institutions' critical role in facilitating development by ensuring transparency in budgeting, metropolitan spending and urban health issues at all levels. These institutions are cornerstones for distributional justice as they facilitate the equitable distribution of resources, as highlighted by Soja (2010). In this way, health resources and services are distributed equitably through effective, accountable and transparent ways. However, the effectiveness of the MMM has been compromised by misgovernance and limited public accountability. Due to this misgovernance, in December 2019, the Free State Provincial Executive used the provisions of section 139(5) (a) and (c) of the constitution of South Africa to place MMM under a mandatory intervention. Specifically, section 139 of the Constitution requires the provincial government to impose a recovery plan to secure the municipality's ability to meet its obligation, which may also lead to the dissolution of a municipality (Republic of South Africa 1996).

However, it has been argued that the fiscal pressure experienced in the MMM has been primarily attributed to the increased pressure placed on the metro after the re-establishment of the metro in 2016, which increased the geographical expansion of the metro to accommodate parts of Naledi Municipality resulting in a need for the service provision to a larger population (Lenka et al. 2021). According to the Auditor General of South Africa (AGSA) (2018), the amalgamation of MMM with other neighbouring small towns coincided with Mangaung having to write off more than R62 million in 2017/18. Moreover, the governance issues in MMM can be attributed to politics of difference inherent in how the African National Congress (ANC) has been experiencing some factionalism that has resulted in the council passing a vote of no confidence in the mayor in August 2020.

The MMM was eventually placed under national administration on 6 April 2022 following the failure to address the issues raised in the previous years resulting from maladministration and poor service delivery. The decision to place the MMM under national administration was made in terms of section 139(7) of the constitution. The major issues concerning the placement of the MMM in national administration include irregular audits that were increasingly being found for the metro. Specifically, the budgetary issues experienced included poor monitoring of internal controls, which have been associated with institutional problems in the metro. For example, it was found that "*Politicians interfere in the administration and specifically in staff appointments and procurement*". Political interference is a critical issue that affects the effective functioning of democratic institutions. It is a factor that derails social justice since inefficiencies are associated with politics of difference, which largely explains the neglect of certain areas in

Bloemfontein regarding accessing basic services as politicians have not been prioritising service delivery (Mphambukeli 2019). On the other hand, politicians do not prioritise critical services such as sewerage treatment plants, which have become hotspots for waterborne disease outbreaks such as cholera and typhoid.

9.6.4 Corruption and Embezzlement of Public Funds

Over the past decade, MMM has been riddled with multiple corruption and gross misuse of public funds. For example, the *Mangaung Metropolitan Municipality Status Quo Report of August 2023* reported rampant unauthorised, irregular, fruitless, and wasteful expenditure in the MMM (MMM 2023). Specifically, as of June 2022, it was indicated that unauthorised expenditure amounted to R5.6 billion, irregular expenditure totalled R3.1 billion, while fruitless and wasteful expenditures were R240 million (MMM 2023: 34). The issue of accountability was also highlighted through the ineffectiveness of the municipality reporting system structure on revenue generated from enforcement of by-laws. In this regard, the municipality loses significant monies that could have been used for several development projects. Moreover, the embezzlement of public funds significantly compromises the development of capital infrastructure, such as clinics and hospitals, which would contribute to healthy cities. This infrastructure deficit has been noted in the informal settlements that lack access to primary healthcare facilities and services, much to the detriment of the urban poor, who spend more on access services (Matamanda et al. 2022).

9.7 Discussion and Policy Implications

The discussion and policy implications of SDG 16 and its impact on health governance and health inequalities are vast and complex (WHO 2018). However, some key points can be highlighted to guide policy and decision-making processes in South Africa.

It is essential to recognise that achieving SDG 16 is not only a responsibility of governments but also of civil society and communities (OECD 2019). All stakeholders must work together to uphold the principles of peace, justice, and inclusion. This requires a collaborative approach that prioritises the needs and voices of communities, especially those marginalised and vulnerable. SDG 16 is an essential component of the broader sustainable development agenda. Achieving other health, education, and economic development SDGs will be difficult without peaceful, just, and inclusive societies. Therefore, policy and decision-makers must integrate SDG 16 into their planning and implementation processes.

Institutional and governance systems in Mangaung impact SD16 in achieving health governance. Socio-spatial justice theory acknowledges the role of institutions in the equitable distribution of resources. Further, spatial injustices hinder citizens and marginalised groups from participating in decision-making processes at the grassroots level. Hence, institutions need to be strengthened to provide access to justice for all, advocating for social justice and inclusivity.

There is a need to strengthen institutions at all governance levels, including local government, to ensure effective service delivery and accountability (OECD 2016). In South Africa, urban areas often face challenges related to poor service delivery and weak governance, leading to inadequate access to basic services such as water, sanitation, and healthcare (Masiya et al. 2019). To address these challenges, policy and decision-makers should prioritise the development of institutional capacity and accountability mechanisms. Addressing violence and insecurity is critical for improving health outcomes, especially in urban areas. Policy and decision-makers should prioritise the development of comprehensive and evidence-based strategies that address the root causes of violence and promote community safety. This requires a multi-sectoral approach that involves collaboration between government, civil society, and communities (Republic of South Africa 1996).

Promoting inclusive decision-making processes and access to justice ensures that all communities have equal access to essential services and can participate in decisions that impact their lives. Policy and decision-makers should prioritise the development of mechanisms that ensure that the voices of marginalised and vulnerable communities are heard and considered (United Nations 2009).

The chapter aimed to explore health governance and inequalities in Mangaung, examined through the intersection of social justice and spatial dimension. Research findings suggest that institutional and governance systems are essential in achieving SDG 16, critical for improving urban health outcomes. As such, there is a need to strengthen institution capacity and accountability mechanisms at all governance levels. In addition, the findings highlight that equitable distribution of resources, opportunities, and spatial is critical to promoting peaceful, just and inclusive societies.

References

Amin A, Graham S (eds) (1997) The contested countryside: rural politics and land controversy in modern Britain. Psychology Press

Auditor General of South Africa (2018) Mangaung Metropolitan Municipality consolidated annual financial statements for the year ended 30 June 2017. Auditor General of South Africa, Pretoria

Baum F, Fisher M, Lawless A (2016) From 'sustainability' to 'resilience': can health inequity be reframed as a wicked problem? Implications for policy change. Soc Sci Med 150:264–271

Cerdá M, Tracy M, Keyes KM (2018) Reducing urban violence: a contrast of public health and criminal justice approaches. Epidemiology 29(1):142–150

Chabalala J (2021) 19 people arrested in Mangaung shutdown protests. News24, 18 May. https://www.news24.com/news24/southafrica/news/19-people-arrested-in-mangaung-shutdown-protests-20210518 (Accessed 21 Apr 2024)

Chapman P, de Langen M (2021) SDG 16 and the 2021 voluntary national reviews: an opportunity to advance justice for all. IISD SDG Knowledge Hub, 7 April. Available from https://sdg.iisd.org/commentary/guest-articles/sdg-16-and-the-2021-voluntary-nationalreviews-an-opportunity-to-advance-justice-for-all/ (Accessed 26 August 2024)

Clarke M (2023) Disgraceful Nehawu shutdown endangering lives. DA, 08 March. https://www.da.org.za/2023/03/disgraceful-nehawu-shutdown-is-endangering-lives (Accessed 23 Apr 2024)

De Villiers K (2021) Bridging the health inequality gap: an examination of South Africa's social innovation in health landscape. Infect Dis Poverty 10(1):19. https://doi.org/10.1186/s40249-021-00804-9

Denters E, Ginther K, de Waart, P. J. (Eds.). (2023) Sustainable development and good governance. Martinus Nijhoff Publishers

Eilertsen A, Pirlea AF, Wang E (2020) The essential role of resilient and responsive institutions. In: Pirlea AF, Serajuddin U, Wadhwa D, Welch M, Whitby A (eds) Atlas of the sustainable development goals 2020: from world development indicators. World Bank, Washington, DC. Retrieved from https://datatopics.worldbank.org/sdgatlas/goal-16-peace-justice-and-strong-institutions/. License: Creative Commons Attribution CC BY 3.0 IGO

Friel S, Akerman M, Hancock T, Kumaresan J (2011) Addressing the social determinants of health: the urban dimension and the role of local government. J Urban Health 88(5):865–874

Government of South Africa (2013) National Development Plan. Government of South Africa, Pretoria

Government of South Africa (2019a) South Africa voluntary national review report. Retrieved from https://sustainabledevelopment.un.org/content/documents/23402SOUTH_AFRICA_RSA_Voluntary_National_Review_Report_Final__14_June_2019.pdf

Government of South Africa (2019b) National health insurance bill. Retrieved from https://www.gov.za/sites/default/files/nationalhealthinsurance.pdf

Government of South Africa (2020) National anti-corruption strategy 2020–2030. Retrieved from https://www.gov.za/sites/default/files/gcis_document/202105/national-anti-corruption-strategy-2020-2030.pdf

Graziano T, Ruggiero L (2023) From periphery to growth pole (and back again?): late industrialism, smart strategies and tourism in South-Eastern Sicily. Reg Stud Reg Sci 10(1):89–105

GroundUp (2023) People's lives are 'not our responsibility' says Nehawu leader as hospital strike intensifies. Mail & Guardian, 9 March. https://mg.co.za/news/2023-03-09-peoples-lives-are-not-our-responsibility-says-nehawu-leader-as-hospital-strike-intensifies/ (Accessed 23 Apr 2024)

Harvey D (1973) Social justice and the city. Johns Hopkins University Press

IFPRI (2019) South Africa: milestones in achieving sustainable development goals for poverty and hunger. [online] IFPRI. Available at: https://www.ifpri.org/publication/south-africa-milestones-achieving-sustainable-development-goals-poverty-and-hunger (Accessed 10 May 2023)

Kruk ME, Gage AD, Arsenault C, Jordan K, Leslie HH, Roder-DeWan S, Adeyi O, Barker P, Daelmans B, Doubova SV, English M, Elorrio EG, Guanais F, Gureje O, Hirschhorn LR, Jiang L, Kelley E, Lemango ET, Liljestrand J, Pate M (2018) High-quality health systems in the Sustainable Development Goals era: time for a revolution. Lancet Glob Health 6:e1196. https://doi.org/10.1016/S2214-109X(18)30386-3

Kühn M (2014) Peripheralisation: Theoretical concepts explaining socio-spatial inequalities. Eur Plan Stud 23(2):367–378. https://doi.org/10.1080/09654313.2013.862518

Lee ACK, Maheswaran R (2011) The health benefits of urban green spaces: a review of the evidence. J Public Health 33(2):212–222

Lenka M, Marais L, Rani K, Tsotetsi LD (2021) Mangaung State of the Cities report—South African Cities Network: Johannesburg

Lima V, Gomez M (2021) Access to justice: promoting the legal system as a human right. In: Leal Filho W, Azul AM, Brandli L, Salvia AL, Ozuyar PG, Wall T (eds) Peace, justice and strong institutions. Springer International Publishing, Cham, pp 1–11

Maphumulo WT, Bhengu BR (2019) Challenges of quality improvement in the healthcare of South Africa post-apartheid: a critical review. Curationis 42(1):e1–e9. https://doi.org/10.4102/curationis.v42i1.1901

Masiya T, Davids YD, Mangai MS (2019) Assessing service delivery: public perception of municipal service delivery in South Africa. Theor Empir Res Urban Manag 14(2):20–40. Retrieved from https://www.jstor.org/stable/26633007

Matamanda AR, Chirisa I, Rammile S, Marais M (2022) Housing and technology: special focus on Zimbabwe (Vol. 37). Springer Nature, Cham

MMM (2023) Mangaung Metropolitan Municipality Status Quo Assessment August 2023. MMM, Bloemfontein

Mochoari R (2021) Free State service delivery protests stall vaccinations in Mangaung. Daily Maverick, 19 May. https://www.dailymaverick.co.za/article/2021-05-19-free-state-service-delivery-protests-stall--vaccinations-in-mangaung/ (Accessed 21 Apr 2024)

Mphambukeli TN (2019) Social justice in planning: Access to adequate drinking water and contested belonging in emerging communities of Mangaung in post-apartheid South Africa. African Renaissance, 16(Special 2), 51–69

Mthethwa A (2021) 'Gatvol' in Mangaung: empty promises fuelled deadly shutdown protest. Daily Maverick, 21 May. https://www.dailymaverick.co.za/article/2021-05-21-gatvol-in-mangaung-empty-promises-fuelled-deadly-shutdown-protest/ (Accessed 21 Apr 2024)

National Academies of Sciences, Engineering, and Medicine (2017) Communities in action: pathways to health equity. The National Academies Press, Washington, DC. https://doi.org/10.17226/24624

Nkuyane L (2022) #MangaungShutdown: residents expected to protest. Ofm, 19 January 2022. https://www.ofm.co.za/article/centralsa/312707/-mangaungshutdown-residents-expected-to-protest (Accessed 21 Apr 2024)

OECD (2016) Governance Notebook 2.6: smoke and mirrors – uncovering the truth about governance. Retrieved from https://www.oecd.org/dac/accountable-effective-institutions/Governance%20Notebook%202.6%20Smoke.pdf

OECD (2017) Illicit financial flows from developing countries: measuring OECD responses. Retrieved from https://www.oecd.org/corruption/illicit_financial_flows_from_developing_countries.pdf

OECD (2019) Achieving the impossible: can we be SDG16 believers? [PDF]. Retrieved from https://www.oecd.org/dac/accountable-effective-institutions/Achieving%20the%20Impossible%20can%20we%20be%20SDG16%20believers.pdf

Omri A, Mabrouk NB (2020) Good governance for sustainable development goals: getting ahead of the pack or falling behind? Environ Impact Assess Rev 83:106388

Republic of South Africa (1996) National crime prevention strategy: summary. Retrieved from https://www.gov.za/documents/national-crime-prevention-strategy-summary

Ruijsbroek A, Droomers M, Groenewegen PP (2017) Hard to reach groups in preventive primary care interventions: a systematic review. Int J Equity Health 16(1):147

Schmid C (2012) Justice and space. In: Sheppard E, Leitner L (eds) The companion to spatial planning. Routledge, pp 47–61

Sclar ED, Garau P, Carolini G, Steel, D. (Eds.). (2005) Urban health: combating disparities with local data. Oxford University Press

SDG Index and Dashboards – Global Data Platform (2022) South Africa SDG Index and Dashboard Report 2022. [online] Available at: https://dashboards.sdgindex.org/static/profiles/pdfs/SDR-2022-south-africa.pdf (Accessed 10 May 2023)

Shulla K, Voigt B-F, Cibian S, Scandone G, Martinez E, Nelkovski F, Salehi P (2021) Effects of COVID-19 on the Sustainable Development Goals (SDGs). Discov Sustain 2(1):15. https://doi.org/10.1007/s43621-021-00026-x

Sifile L (2021) The sparks that were fanned into flames of Bloem protests. Sowetanlive, 24 May. https://www.sowetanlive.co.za/news/south-africa/2021-05-24-the-sparks-that-were-fanned-into-flames-of-bloem-protests/ (Accessed 21 Apr 2024)

Soja E (2010) Seeking spatial justice. University of Minnesota Press

Sommerlad H (2004) Some reflections on the relationship between citizenship, access to justice, and the reform of legal aid. J Law Soc 31(3):345–368

South African Sustainable Development Goals Hub (n.d.) Home. [online] Available at: https://sasdghub.up.ac.za/home/ (Accessed 10 May 2023)

Statistics South Africa (2019) SDGs Country Report 2019: South Africa [PDF]. Retrieved from http://www.

statssa.gov.za/MDG/SDGs_Country_Report_2019_South_Africa.pdf
StatsSA (2019) Sustainable development goals: country report 2019
Swyngedouw E (2004) Social power and the urbanisation of water: flows of power. Oxford University Press
United Nations (2009) Towards an inclusive society: The case of Ghana. Retrieved from https://www.un.org/esa/socdev/egms/docs/2009/Ghana/inclusive-society.pdf
United Nations (2015) Transforming our world: the 2030 Agenda for Sustainable Development. Retrieved from https://sdgs.un.org/2030agenda
United Nations (2017) Official List of Proposed SDG Indicators. Retrieved from https://sustainabledevelopment.un.org/content/documents/11803Official-List-of-Proposed-SDG-Indicators.pdf
UNSTATS (2020) The Sustainable Development Goals Report 2020. United Nations. https://unstats.un.org/sdgs/report/2020/The-Sustainable-Development-Goals-Report-2020.pdf
WHO (2017) Regional action plan on health promotion in the sustainable development goals (2018–2030). Retrieved from https://www.who.int/publications/i/item/9789290226383
WHO (2018) Health inequities and their causes. Retrieved from https://www.who.int/news-room/facts-in-pictures/detail/health-inequities-and-their-causes
Zerbo A, Delgado RC, González PA (2020) Vulnerability and everyday health risks of urban informal settlements in Sub-Saharan Africa. Glob Health J 4(2):46–50. https://doi.org/10.1016/j.glohj.2020.04.003

Lived Experiences of the Energy Consumers of Selected Sites in Disadvantaged Communities in Bloemfontein, South Africa

10

Kgosi Mocwagae, Innocent Chirisa, and Tebello Putsuoane

10.1 Introduction

South Africa has been faced with energy production challenges in the form of electricity from ESKOM (Electricity Commission) since 2007, leading to households having to rely on alternative energy sources. As commonly known, energy is required for cooking, heating, cooling, lighting powering machines through physical and chemical resources. Through all these challenges, Bloemfontein has not been spared because its households also rely on the same troubled electricity source for energy. The study has particularly focused on low-income households in disadvantaged communities in Bloemfontein as they generally will have affordability issues in terms of accessing low-carbon emission energy alternatives.

Hereafter, the chapter first reviews current energy-related knowledge such as load shedding, renewable and non-renewable energy sources, and low-income energy options. Second, the chapter has a research methodology that outlines the research design, data collection tools, sampling, methods of analysis, and ethical considerations. Third, results and discussion are presented on data collected from the study area to gather facts on the access to energy encountered by households. Lastly, the is a conclusion and discussion of the study to propose ways through which income low-income households of the study area can best access energy.

10.2 Literature Review

The literature review section includes four sections that unpack literature related to energy access in South Africa and globally. The first section unpacks what load shedding is and its impact on communities. The second section how informal settlement dwelling in light of sustainable goals 7 and 12. The third section is on renewable and non-renewable fuels as energy sources. The fourth section is on low-income energy consumption to foster sustainable development.

K. Mocwagae (✉) · T. Putsuoane
Department of Urban and Regional Planning, University of the Free State, Bloemfontein, South Africa
e-mail: MocwagaeKS@ufs.ac.za

I. Chirisa
Department of Urban and Regional Planning, University of the Free State, Bloemfontein, South Africa

Office of the Vice Chancellor, Zimbabwe Ezekiel Guti University, Bindura, Zimbabwe

A. R. Matamanda, V. Nel (eds.), *Sustainable Development Goals and Urban Health*, Sustainable Development Goals Series, https://doi.org/10.1007/978-3-031-68734-1_10

10.2.1 Understanding Load Shedding and Its Impact on Communities

Load shedding is an electricity management practice that is utilised to prevent an electricity grid collapse or blackouts. Taylor (1992) refers to it as undervoltage load shedding which became a partial solution to voltage stability challenges in the Northeast in 1965. Load shedding occurs when there is a huge demand for electricity versus available supply (Masinga and Madzivhandila 2023). Load shedding has been defined as an electrical power outage, as the electrical power system is forced to shut down, allowing some areas to receive electricity while others are temporarily cut out this happens when the power supply can no longer meet the entire demand (Ibrahim et al. 2023). For Lakra and Kirar (2015), load shedding is done to address the gap between power generation and demand which requires adequate techniques for fast and reliable shedding to avoid blackouts. Blackouts occur when the demand for electricity exceeds the supply and cause an imbalance that causes a trip of the system entirely (Eskom 2023). In the case of a blackout, it would take up to two weeks to restore the electricity at the plants to restore full power. Therefore, loadshedding as an electricity management practice has good intentions, however, results in negative effects to low income households.

Masinga and Madzivhandila (2023) observes that during apartheid in South Africa, electricity was reserved or designated to benefit minority population, predominantly of white descent however, this changed when South Africa, transitioned into a rainbow nation or a state of new dispensation. The previously excluded ethnic groups (black) were to be accommodated into the same electricity grid despite the urgent need for inclusion, government institutional bodies failed to carter to the growing demand this failure has since escalated into load shedding (Masinga and Madzivhandila 2023). The ripple effects of black outs and load shedding became a South African phenomenon intertwined with the lived experiences of South African and load shedding will continue in future as electricity generation infrastructure is approaching a time lapse in which it will need to be replaced (Mutambo et al. 2023; Meyer and Neethling 2023).

The track record off South Africa trying to build new electricity generation capacity has not been successful since the load shedding crisis began in 2007 (Wernecke et al. 2024). The year 2022 was set to become the worst year of load shedding on record according to Eskom as of October, 8 the country had experienced 1949 hours, or 81 days of national load shedding in 2022 this is almost double that of the 1153 hours, over 48 days in 2021 (Lesala et al. 2023; Majola and Langerman 2023; Mukumba and Chivanga, 2023). Without access to electricity this creates a challenges because most of the disadvantaged people especially in urban informal settlements rely on it for keeping food properly stored and refrigerated, among other (Monyai et al. 2023). Informal residents have a high demand for energy to an extent that they connect their households to the electrical grid through illegal means (Monyai et al. 2023).

10.2.2 Informal Settlement Dwelling in Light of Sustainable Development Goals 7 and 12

The mounting pressure in urban areas for people to find places to live results in challenges that include informal settlements. These informal settlements are typically affordable and well-located settlements close to employment opportunities (Turok 2015). Unfortunately, informal settlements are at times situated in dangerous environments due to a lack of technical insights needed in their establishment (Satterthwaite et al. 2020). The UN-Habitat (2015) indicates that the emergence of informal settlements is caused by different factors such as rural-urban migration, population growth within cities, poor planning governance, high housing prices, and economic disadvantages. With all these associated dynamics of the emergence of informal settlements, they remain economically, geographically, socially, and politically separated from local governance procedures related to development opportunities and decision-making (Avis 2016).

10.2.3 An Evaluation of Renewable vs Non-Renewable Fuels

Energy is accessible through renewable and non-renewable energy sources. Non-renewable, commonly known as fossil fuels, are energy sources that are the most commonly utilised, more so in the Global South (Goldthau et al. 2020). Fossil fuels such as oil, natural gas and coal are a result of decomposed carbon-based organisms that died millions of years ago and are perishable after use. According to Abas et al. (2015), fossil fuel reserves are steadily increasing without there being a sign of depletion in the near future even though they are finite resources. Fossil fuels are a frequent global point of discussion as far as energy provision because of the negative environmental effects they are associated with. From 2011, fossil fuels accounted for approximately 83% of energy production (Perera 2018), with an ambitious target of 0% fossil fuel energy production by 2050 according to Holechek et al. (2022).

Based on the target set for 0% fossil fuel energy production, this has put a challenge for renewable energy sources to be explored and developed through a process referred to as a Just Energy Transition (JET) (Suboticki et al. 2023; Sun et al. 2023). In order to achieve JET, Lin and Li (2022) emphasise that there is a growing need for energy while trying to combat carbon emissions through clean energy through renewable energy sources. The implications of clean energy provision through renewable energy sources affect regional pollution emissions, energy supply structures, energy system costs, and energy security. According to the United Nations (2023), renewable energy is derived from natural sources that are produced as a higher rate than they are consumed and include six sources such as solar, wind, hydro, geothermal, ocean and biomass energy.

There are advantages and disadvantages when considering renewable energy sources. There are advantages to considering renewable energy sources (Ploetz et al. 2016; Maradin 2021; Mohtasham 2015). The first is that they do not produce carbon emissions from fossil fuels and reduce air pollution. Second, is the diversification of energy supply and reducing reliance on imported fuels. Third, it creates economic development and jobs in installation, manufacturing and related activities. It should however be noted that there are six challenges with renewable energy sources. The first challenge is that the initial costs of implementation are higher than fossil fuels, with solar and wind energy being the more affordable options. Second, lack of infrastructure because fossil fuel infrastructure is not capable of handling the amount of renewable energy sources. Third, renewable energy sources require power storage unlike fossil fuels which distribute energy immediately after generation. Fourth, the monopoly by major fossil fuel companies has major support with the government and while renewable energy suppliers do not enjoy the same benefits. Fifth, local communities have a lack of knowledge on the access of renewable energy sources compared to fossil fuels. Last, there is a lack of incentives and policy frameworks to support the adoption of renewable energy sources.

Sustainable Development Goals (SDGs) 7 and 12 are the United Nations policy response towards sustainable access, consumption and production of energy sources by 2030. SDG 7 aims to ensure access to affordable, reliable, sustainable and modern energy for all. Target 7.1 focusing on ensuring universal access to affordable, reliable and modern energy services. By achieving target 7.1, this will result in low-income households in informal settlements and will make it more accessible as renewable energy is expensive (Lup et al. 2023; Chirambo 2018). SDG 12 aims to ensure sustainable consumption and production patterns. Target 12.2 aims to achieve sustainable management and efficient use of natural resources, primarily focusing on the reducing the reliance on fossil fuels and investing in renewable energy sources instead (Fuso Nerini et al. 2018; Alola et al. 2021).

10.2.4 Low-Income Energy Consumption

Energy consumption in low-income households is generally affected by economic factors that result in negative environmental and social impacts. To better define the economic challenges encountered by low-income households, the term energy poverty is used globally which refers to a lack of access to electricity and/or inability to afford access to energy that one needs (Ugarte et al. 2016; World Bank 1999; Murshed and Ozturk 2023). Furthermore, this section intends to briefly discuss the types of energy sources utilised by those that experience energy poverty.

According to Ugarte et al. (2016) there are six barriers to affecting low-income households. The first barrier is behavioural which is based on a lack of awareness of benefits associated with energy efficiency measures. The second barrier is a lack of informational dwelling consumption, lack of knowledge of consumption, and no understanding of the difference between energetic improvements due to new investments and general maintenance. The third barrier is a lack of understandable information and support programmes that provide grants. The fourth barrier is economically related to internal household factors such as a lack of savings and external household factors due to a lack of access to credit. The fifth is a lack of financial incentives because of a lack of land ownership which can be entered into by energy producers and subsidies on energy prices on energy they produce. The last barrier is a lack of risk aversion because of the time required for amortisation, uncertainty of the future, and a preference for equality other than debt.

10.2.5 Negative Impact of Energy Production and Consumption on Health

The impact of energy production and consumption cannot be excluded from the discussion of energy access for low-income earners of the selected settlements in Bloemfontein. Energy production and consumption have various negatives which are social, economic, and environmental, and this section focuses on health as a social impact (Markandya and Wilkinson 2007). The negative impact of health burden on communities as already stated in Sects. 10.2.3 and 10.2.4 is a cause of greenhouse gases (CO2). National Research Council, Policy, Global Affairs, Board on Science, Economic Policy, Division on Engineering, Physical Sciences, Board on Energy, Environmental Systems, Division on Earth and Life Studies (2010) and Markandya and Wilkinson (2007) indicate that different natural resources have different negative impacts on health. Coal and oil are considered the higher emitters of greenhouse gases as compared to natural gas and nuclear power (Kwakwa et al. 2020; Markandya and Wilkinson 2007).

The results and health burden consequences to humans are linked to conditions such as bone atrophy, obesity, respiratory diseases, chronic inflammation, cancer, kidney failure, and loss of brain function (Bierwirth 2018; Duarte et al. 2020). The health burdens can therefore be flagged as causing cancerous and/or neurological disorders among others. The challenges of primary healthcare access to low-income earners in South Africa are well-documented as they are dependent on the state due to the non-affordability of the better-performing private healthcare system (Maphumulo and Bhengu 2019; Malakoane et al. 2020; Burger and Christian 2020). Based on the associated health burden caused by greenhouse gases, it is of great importance to reduce the vulnerability of low-income earners in their effort to access energy.

10.3 Research Methodology

Methodologically, this paper employed a qualitative approach using a phenomenological framework. This approach was used to understand the "meaning of life" for informal settlements in Bloemfontein. The Four informal settlements that were purposefully sampled are: Phomolong, Bochabela, Phase 2, and Turflaagte. These four informal settlements were selected because they provide a variety of new and old human settlements in the Mangaung Metropolitan Municipality. Phomolong, the most recent informal settlement among the selected study areas, was established during COVID-19 to fight the pandemic by decongesting other informal settlements. The informal settlement is four years old while it is inhabited by residents from other informal settlements and neighboring areas. This settlement exhibits characteristics of informal settlements in South Africa with the lack of essential services, poverty stricken, temporary housing structures, social distress, and a high unemployment rate (Ziblim et al. 2013). Likewise, these were the distinctive characteristics of Bochabela, Phase 2, and Turflaagte.

Data for the study has been collected from primary sources. Data was retrieved through semi-structured interviews, questionnaires, observations, and transect walks for each informal settlement. These data sources depicted how people in informal settlements give meaning to their daily lives. A sample population of forty respondents were selected and interviewed in each informal settlement, with their responses discussed in the section below.

Table 10.1 South African load shedding stages

Stages	Reduced electricity consumption	Duration
1	1000 MW	Three two-hour outages over four days or three four-hour outages over eight days
2	2000 MW	At least six times for four hours over eight days or six times over four days for two hours at a time
3	3000 MW	Consists of nine times over eight days for fours or nine times for two hours with no power for four days
4	4000 MW	Twelve four-hour outages over eight days or twelve two-hour outages over four days
5	5000 MW	At least four days for twelve periods of two hours
6	6000 MW	Unexpected power cuts outside load shedding schedules. It occurs eighteen times over four days for 4½ hours at a time or eighteen times over eight days for 2 hours at a time.
7	7000 MW	Four-hour slots for four days
8	8000 MW	No power six times a day for twelve hours daily

Source: Ecoflow (2023)

10.4 Results

10.4.1 Loadshedding in South Africa and Its Impact on Bloemfontein

Load Shedding has become a standard practice in South Africa since 2007 (Goldberg 2015), and has since been utilised based on different events to date. The implementation of load shedding in South Africa is done in eight different stages (Table 10.1) with time schedules that represent reduced electricity consumption in megawatts (MW) (Ecoflow 2023).

As much as load shedding has been stated to be advantageous in terms of mitigating blackouts, unfortunately, the struggle with load shedding in South Africa has had negative effects on communities and the economy (Mlambo 2023). Business Tech (2024) observes that the latest edition of the code introduced some changes to the load shedding structure in the country, the most notable of which is the extension of load shedding stages to stage 16 in the previous editions of the code, load shedding was only structured to stage 4 in the first edition and then to stage 8 in the second, showing the exact schedule and time-periods attached to rolling blackouts for 25% and then 50% of the total load, respectively. The resultant effect of a lack of electricity supply is the inability of the Mangaung Metropolitan Municipality to be able to formalise informal settlements. Business Tech (2024) indicates that one thing that has changed in the new edition, which may impact end-users and customers, is that the load shedding cycles have been altered with the third edition schedules still following 32-hour cycles.

Malope (2022) alludes that because of load shedding 93% of South Africans living in Bloemfontein and Free State have been forced to throw away food which was spoilt in refrigerators due to the damage caused by power outages. Inclusive Cities (2023) indicates that South Africa is facing a major energy crisis, characterised by widespread load shedding in Free State Province this has been worse and this has a significant impact on the urban economy and cities. Inclusive Cities (2023) observes that the impacts of the energy crisis on the urban economy has been far-reaching and had major consequences for business, households, and individuals thereby affecting everyone as urban economies are complex and interconnected.

10.4.2 Informal Settlement Dwelling in Bloemfontein

South Africa, even though often presented as if it only occurred post-apartheid, has had to deal with informal settlements since colonial times (Brown-Luthango et al. 2017). Furthermore, the rationale for relocation to informal settlements is access to socioeconomic opportunities, living closer to family members, and access to basic services. Jones (2017) emphasises that the responses to informal settlements by institutions and government legislative frameworks have to promote human rights and provide adequate housing opportunities. The National Upgrading Support Programme (NUSP) was formulated as a policy response by the Department of Human Settlements to address informal settlements (Brown-Luthango et al. 2017). The informal settlements in Bloemfontein will thus need to be responded to using the NUSP.

10.4.3 Lived Experiences of Energy Access by Low-Income Settlement Dwellers in Bloemfontein

The lived experiences of energy access by low-income settlement dwellers in Bloemfontein were conducted in Phase 2, Bochabela and Phomolong. These settlements were chosen because they form part of the areas in Bloemfontein with low-income settlers that are dependent on social welfare programmes of government for their livelihoods. Results in this section as discussed into sections, first for Phase 2 where energy access data was collected, and second for Bochabela and Phomolong where some responses were given but not as detailed as Phase 2.

10.4.3.1 Responses from Phase 2

There were 10 household participants from Phase 2 who participated in the study to respond specifically to energy.

10.4.3.1.1 Demographic Profile of Participants

All the participants of the study identified as black. As shown in Table 10.2, eight of the 10 participants were female and two were male, suggesting that most of them could be female-headed households.

As shown in Table 10.3, most of the participants (4/10) were between 30–39 years old, followed by those between 40–49 years old (3/10), 60 years and older (2/10), and 50–59 years (1/10). Based on the ages of the participants, this demonstrates that low-income earners are more in the younger years and decrease the older they grow.

Table 10.4 confirms that most of the participants (6/10) are single, followed by those who are married (2/10), with one being widowed and another with a life partner. Having a majority of participants being single reduces the prospects of increasing the disposable household income.

The level of education of participants as depicted in Table 10.5 indicated that a majority of participants have some high school education

Table 10.2 Gender of participants from Phase 2, Bloemfontein

Variables	Frequency
Male	2
Female	8
Total	10

Source: Author (2024)

Table 10.3 Age of Participants from Phase 2, Bloemfontein

Variables	Frequency
20–29	0
30–39	4
40–49	3
50–59	1
60+	2
Total	10

Source: Author (2024)

Table 10.4 Marital status of participants from Phase 2

Variables	Frequency
Single	6
Married	2
Life partner	1
Widowed	1
Total	10

Source: Author (2024)

(4/10), followed by those who have primary schooling (3/10), and three who each have matric (1/10), diploma (1/10) and a technical certificate (1/10). Based on the level of education of the majority of participants from Phase 2, suggests that employment prospects for high-paying work are significantly reduced.

In Table 10.6, there are only three participants who are employed and one self-employed participant. However, three participants are unemployed, and three are retired and have a reduced earning capacity.

In Table 10.7, the household size of participants in Phase 2 was also collected. A majority of households (5/10) have three household members, followed by two household members (3/10), one living by themselves and another one that has nine household members.

10.4.3.1.2 Energy Access in Phase 2

Phase 2 is in a formally established township with access to electricity, despite its affordability decreasing due to increasing costs. Nine participants stated that they used paraffin as a source of energy for heating and cooking. Only one of the participants indicated that they use candles and

Table 10.5 Level of education of Phase 2 participants, Bloemfontein

Variables	Frequency
Primary schooling	3
High schooling	4
Matric	1
Diploma	1
Technical certificate	1
Total	10

Source: Author (2024)

Table 10.6 Employment status of Phase 2 participants

Variables	Frequency
Unemployed	3
Employed	3
Self-employed	1
Retired	3
Total	10

Source: Author (2024)

Table 10.7 Household size of participants in Phase 2

Variables	Frequency
1	1
2	3
3	5
4	0
5+	1
Total	10

Source: Author (2024)

gas as a source of energy. Gas was mentioned by two participants as a reliable energy source for heating and cooking, but the cost and time of accessing was indicated to not be as convenient as electricity.

10.4.3.1.3 Consideration of Alternative Energy Sources

When Phase 2 participants were asked what it would take for them to consider alternative and environmentally friendlier energy sources, they had the following responses. All 10 participants indicated that affordability was the barrier for them considering alternative and environmentally friendly energy sources and are left to use electricity despite it being expensive and unreliable at times. The primary challenge they indicated reduces their affordability is due to rising transport, food, and living costs in general.

10.4.3.2 Responses from Bochabela and Phomolong

Responses from Bochabela and Phomolong were grouped they also encountered challenges for energy access as low-income earners.

10.4.3.2.1 Availability of Electricity

All the respondents in Bochabela, and Turflaagte indicated that they have access to electricity. Surprisingly, when the residents of Phomolong were asked "over the past one year, how often, if ever, have you or your family gone without electricity?" all the responses were "always". They said the councilor has been making promises that electricity will be installed soon. Figure 10.1 shows a prima stove and a makeshift one burner stove which are the assets used for food preparation in Phomolong. None of the respondents knew the consequences of using fossil fuels for energy consumption. Given that there is not electricity, the respondents indicated that canning bottles are used to preserve fruits and food is preserved by putting it on the floor.

Figure 10.2 indicates the solar panels used by a significant number of households in Phomolong to connect mainly the TVs and radios, but to occasionally light the house. The respondents said that they resorted to using this clean energy because it is more affordable than gas and paraffin. Due to the high demand of gas and paraffin, these coping strategies that have been indicated encouraged other residents to informal sell gas in the neighborhood (Fig. 10.3).

10.4.3.2.2 Impact of Load Shedding

Load shedding has become progressively dreadful post-Covid. Settlements that have electricity such as Phase 2, Bochabela, Turflaagte experience stages one to six of load shedding. When asked how efficient is the energy supply, one respondent in Phase 2 said:

> it is the worst. There is an app for load shedding schedules, but they no longer apply. Load shedding happens at any time for long periods. It has added onto the household financial costs because food

Fig. 10.1 A prima stove and a make shift one burner stove in Phomolong

Fig. 10.2 Solar panels in Phomolong

> can't be bought that goes in the fridge because of the fear of it spoiling. so things like meat are bought frequently and must be eaten quickly.

Another respondent added:

> it is not efficient because of load-shedding. Load shedding has made daily lives inconsistent like all cooking for the day has to be done early in the morning. When there is load shedding, nothing can be done. All house activities stop.

Figure 10.4 shows that there are wooden utility poles that distribute electricity in Turflaagte, which are also used in Bochabela and Phase 2. Thus, the households have electricity but cannot use it due to impacts of load shedding. This has severely affected the livelihoods of people living in the three settlements. One of the respondents from Turflaagte stated that:

> load shedding has been dire to the carwash because it is the main source of income. People come to this carwash in particular because of the pressure washer. So it cant be used during load shedding and normal hose pipes are used. Some customers feel like their cars still need to be fully washed and they never return.

10.4.3.2.3 Alternative Fuels and Technologies Used

The study found that the settlements that have access to electricity also have to find alternatives when there is load shedding. All the respondents had the same answer indicating their alternatives of choice:

> paraffin. Cooking, heating, and boiling water for bathing are all done on a paraffin heater, even though paraffin is now expensive a litre is now around R28.00.

This is evident that electricity is not a reliable source of energy in Bochabela, and Turflaagte. therefore, they have to compensate with paraffin and gas which are both used for cooking, boiling water for bathing, and for heat. The alternative fuels used are similar to the ones used in Phomolong and they can be seen in Fig. 10.1.

Fig. 10.3 Informal household selling gas

Fig. 10.4 Electricity reticulation in Turflaagte

10.5 Discussion

The lived experiences of low-income earners in Bloemfontein in terms of access to electricity as a primary energy source have some commonalities and differences in certain regards. Residents of Phase 2, Bochabela and Tuurlaagte seemed not to have had issues with access to electricity as they live in formally established areas, whilst the residents of Phomolong which residents were settled on in 2020 seem not to have access to electricity at all. The experience of the residents of Phomolong signifies that with the increase of the urban population in Bloemfontein, the municipality has a challenge meeting their primary energy needs using electricity. Despite promises made by the ward councillor regarding access to electricity, this has not been a reality. This is not concurrent with Masinga and Madzivhandila (2023) who posits that urban areas must provide universal access to affordable, sustainable and modern energy services by 2030 as provided by SDG 7. Ugarte et al. (2016) defined this lack of access to electricity as energy poverty.

However, the residents of Phomolong have found alternative ways to cope without electricity. Residents of Phomolong rely on fossil fuels such as paraffin and liquid petroleum gas for the preparation of their food, candles for lighting and small-scale solar panels for energy generation for charging their cell phones, watching TV and listening radio. Concurrent with the study is Amabuo et al. (2023) that observes that in Africa there has been the use of solar energy to provide cost-effective and sustainable power to potentially reduce or eliminate the need for diesel generators. Similar to the study are Cheruiyot et al. (2024) that alludes that there is a uniform use of gas for cooking across South Africa followed by candle lighting and solar is a distant third choice.

However, even though Phase 2, Bochabela and Turflaagte residents had access to electricity, they were affected by load shedding. Due to their low-income earning status affordability of renewable and/or alternative energy sources was a challenge, leaving them unable to avert the impact of load shedding on their access to energy. Fossil fuels then became the preferred means of energy access during load shedding which negatively contributes to the increased carbon footprint. The study shows that load shedding experiences are linked to people's socio-economic background similar to these findings is Mukoni (2012) that observes that people's socio-economic background plays a critical role in their experiences of load shedding with people from low-economic backgrounds suffering load shedding the most. Consistent with the study is Musademba et al. (2012) that alludes that in Zimbabwe preferential treatment was given to specific socio-economic classes in terms of hours of load shedding experience with residents of low-income areas experiencing longer hours of load shedding relative to their counterparts in high-income areas. Similar to these findings are Aidoo and Briggs (2019) that observes that rolling blackouts proportionately affect the poor and among the sources of this inequality are low-income households who are believed to be making minimal contribution towards economic outputs in society. Thus, load shedding is based on segregation along the lines of economic output that a community makes in most African countries. Masinga and Madzivhandila (2023) alludes that this narrative of neglecting the poor does not consider that as much as the poor might contribute minimally they are still significant contributors to various spheres of society this narrative poses risk and vulnerability threat, which can reach a point where they do not contribute at all if they are not prioritised to access adequate services such as electricity. This can further increase the gap between the rich and poor with the latter being left behind failing the SDGs as they intend to leave no one behind.

Based on the discussion of the findings, SDGs 7 and 12 still require attention in low-income areas of Bloemfontein. Ensuring that issues of urban health are concerned the study should use SDGs 7 and 12 to drive clean energy and responsible consumption and production of reduced CO2 energy sources. The primary goal of SDG 7 to affordable and clean energy still presents challenges because of a combination of lacking state capacity to produce energy and the inability to afford renewable or alternative energy sources which need to be paid attention to for better livelihoods. Without SDG 7 being attended to, SDG 12's aspirations of responsible consumption and production will remain challenged because low-income residents will not be able to explore access to renewable and alternative sources that emit less carbon and consume less of the world's natural resources.

In support of the study are Nduhuura et al. (2020) that observes that the energy crisis has contributed to the wood fuel crisis in the Third World, further resulting in the decrease of forests in African countries resulting in the failure to attain the targets of SDG 7 that targets access to sustainable energy. Similar to the study are Wiese and van der Westhuizen (2024) that access to reliable electricity supply is imperative for progress however, consumers in numerous developing countries such as South Africa do not have access to reliable electricity supply owing to power cuts and power outages compounding the attainment of SDG 7. The inequalities in the access to clean and reliable energy remains the stumbling block to the attainment of SDG 7 in Africa with most of the load shedding experiences modelled around segregatory lines with the poor bearing the burden of longer load shedding schedules. The load shedding experiences are modelled based on community contribution towards the socio-economic contribution to the economy with the poor being marginalised because of their low economic contribution.

10.6 Conclusion and Future Direction

To achieve SDG 7 and 12, disadvantaged communities as the majority of the population in South Africa and Bloemfontein need to be made central in the access to renewable and alternative energy sources. With SDG 7 and 12 being intertwined, this chapter proposes that the government has to introduce policies that will provide incentives for the provision of affordable and clean energy, which allows for responsible consumption and production in disadvantaged communities. Progressive conditions of those who are economically disadvantaged can only be improved by allowing them access to sustainable resources because it reduces the state's spending on corrective negative impacts and also increasing opportunities for employment in such communities. Further research has to be done on how policies can be introduced to allow for disadvantaged communities to have access energy consistent with the SDGs.

References

Abas N, Kalair A, Khan N (2015) Review of fossil fuels and future energy technologies. Futures 69:31–49

Aidoo K, Briggs RC (2019) Underpowered: rolling blackouts in Africa disproportionately hurt the poor. Afr Stud Rev 62(3):112–131

Alola AA, Ozturk I, Bekun FV (2021) Is clean energy prosperity and technological innovation rapidly mitigating sustainable energy-development deficit in selected sub-Saharan Africa? A myth or reality. Energy Policy 158:112520

Amabuo C, Ibrahim B, Ebere E, Ayodele E (2023) The use of solar energy to power the oil and gas FPSO in Africa. In SPE Nigeria Annual International Conference and Exhibition (p. D021S011R005). SPE

Avis W (2016) Urban governance (topic guide). In: Birmingham, UK: GSDRC. University of Birmingham, Birmingham, pp 1–57

Bierwirth P (2018) Carbon dioxide toxicity and climate change: a major unapprehended risk for human health. Web Published: ResearchGate, 10

Brown-Luthango M, Reyes E, Gubevu M (2017) Informal settlement upgrading and safety: experiences from Cape Town, South Africa. J Housing Built Environ 32

Burger R, Christian C (2020) Access to health care in post-apartheid South Africa: availability, affordability, acceptability. Health Econ Policy Law 15(1):43–55

Business Tech (2024) South Africa now has 16 stages of load shedding. [Online] Available at https://businesstech.co.za/news/energy/766453/south-africa-now-has-16-stages-of-load-shedding/ [Accessed on 10 April 2024]

Cheruiyot K, Lengaram E, Siteleki M (2024) South Africa's energy landscape amidst the crisis: unpacking energy sources and drivers with 2022 South African census data. Sustain For 16(2):682

Chirambo D (2018) Towards the achievement of SDG 7 in sub-Saharan Africa: creating synergies between Power Africa, Sustainable Energy for All and climate finance in-order to achieve universal energy access before 2030. Renew Sust Energ Rev 94:600–608

Duarte C, Jaremko Ł, Jaremko M (2020) Hypothesis: potentially systemic impacts of elevated CO2 on the human proteome and health. Front Public Health 8:543322

Ecoflow (2023) What are the stages of load-shedding? [Online] Available at: https://blog.ecoflow.com/za/load-shedding-stages/ (Accessed 22 Aug 2023)

Eskom (2023) What is load shedding. [Online] Available at: https://loadshedding.eskom.co.za/LoadShedding/Description (Accessed 22 Aug 2023

Fuso Nerini F, Tomei J, To, L.S, Bisaga I, Parikh P, Black M, Borrion A, Spataru C, Castán Broto V, Anandarajah G, Milligan B (2018) Mapping synergies and trade-offs between energy and the Sustainable Development Goals. Nat Energy 3(1):10–15

Goldberg A (2015) The economic impact of load shedding: the case of South African retailers. Doctoral dissertation, University of Pretoria, Pretoria

Goldthau A, Eicke L, Weko S (2020) The global energy transition and the global South. Geopol Glob Energy Trans 73:319–339

Holechek JL, Geli HME, Sawalhah MN, Valdez RA (2022) Global assessment: can renewable energy replace fossil fuels by 2050? Sustain For 14:4792. https://doi.org/10.3390/su14084792

Ibrahim HA, Ayomoh MK, Bansal RC, Gitau MN, Yadavalli VS, Naidoo R (2023) Sustainability of power generation for developing economies: A systematic review of power sources mix. Energy Strateg Rev 47:101085

Inclusive Cities (2023) Tri-fold Impact of load-shedding in South African Cities: Power Outage, Destruction, and Regression. [Online] Available at https://inclusivecities.ukzn.ac.za/blog/tri-fold-impact-of-load-shedding-in-south-african-cities-power-outage-destruction-and-regression/ [Access on 16 November 2023]

Jones P (2017) Formalising the informal: understanding the position of informal settlements and slums in sustainable urbanisation policies and strategies in Bandung, Indonesia. Sustain For:9

Kwakwa P, Alhassan H, Adu G (2020) Effect of natural resources extraction on energy consumption and carbon dioxide emission in Ghana. Int J Energy Sect Manag 14(1):20–39

Lakra P, Kirar M (2015) Load shedding techniques for system with cogeneration: a review. Elect Electron Eng: Int J (ELELIJ) 4

Lesala ME, Shambira N, Makaka G, Mukumba P (2023) The energy poverty status of off-grid rural households: a case of the upper Blinkwater Community in the Eastern Cape Province, South Africa. Energies 16(23):7772

Lin B, Li Z (2022) Towards world's low carbon development: the role of clean energy. Appl Energy 307:118160

Lup ANK, Soni V, Keenan B, Son J, Taghartapeh MR, Morato MM, Poya Y, Montañés RM (2023) Sustainable energy technologies for the Global South: challenges and solutions toward achieving SDG 7. Environ Sci: Adv 2(4):570–585

Majola CMD, Langerman KE (2023) Energy efficiency in the South African mining sector: A case study at a coal mine in Mpumalanga. J South Afr Inst Min Metall 123(9):451–462

Malakoane B, Heunis J, Chikobvu P, Kigozi N, Kruger W (2020) Public health system challenges in the Free State, South Africa: a situation appraisal to inform health system strengthening. BMC Health Serv Res 20:1–14

Malope BJ (2022) Power struggles: an exploration of the contribution of renewable energy to sustainable development, decent work and the "just transition" through a case study of wind farm development outside Loeriesfontein, Northern Cape Province (2011–2020). Doctoral dissertation, Stellenbosch University, Stellenbosch

Maphumulo W, Bhengu B (2019) Challenges of quality improvement in the healthcare of South Africa post-apartheid: a critical review. Curationis 42(1):1–9

Maradin D (2021) Advantages and disadvantages of renewable energy sources utilization. Int J Energy Econ Policy 11(3):176–183

Markandya A, Wilkinson P (2007) Electricity generation and health. Lancet 370(9591):979–990

Masinga F, Madzivhandila T (2023) Loadshedding impact on food spoilage: an analysis of household experiences in South Africa. Afr J Govern Dev 12(2):182–197

Meyer DF, Neethling JR (2023) Electricity production and economic growth: the South African case. Afr J Busin Econ Res 18(1):143

Mlambo V (2023) Living in the dark: load shedding and South Africa's quest for inclusive development. IAHRW Int J Soc Sci Rev 11(2):153–160

Mohtasham J (2015) Renewable energies. Energy Procedia 74:1289–1297

Monyai P, Chivanga SC, Katsande N (2023) Inequalities in access to energy in informal settlements: towards energy justice in Gqeberha and Komani in South Africa. Water-Energy Nexus 6:1–5

Mukoni M (2012) The impact of load shedding on gender relations in heterosexual households in Mkoba north, Gweru, Zimbabwe: implications for sustainable development. Int J Dev Sustain 1:402–416

Mukumba P, Chivanga SY (2023) An overview of renewable energy technologies in the eastern Cape Province in South Africa and the rural households' energy poverty coping strategies. Challenges 14(1):19

Murshed M, Ozturk I (2023) Rethinking energy poverty reduction through improving electricity accessibility: a regional analysis on selected African nations. Energy 267:126547

Musademba D, Kanyepe M, Madiye L, Hove T (2012) Effect of load shedding in Chinhoyi urban residential areas, Zimbabwe. Int J Energy Eng 2(5):232–241

Mutambo H, Kawimbe S, Meki-Kombe C, Mwange A (2023) Impact of electricity load shedding on operations of small-scale enterprises in selected developing

countries: a review of literature. Journal of Economics and Sustainable Development 14(13):54
Nduhuura P, Garschagen M, Zerga A (2020) Mapping and spatial analysis of electricity load shedding experiences: a case study of communities in Accra, Ghana. Energies 13(17):4280
Perera F (2018) Pollution from fossil-fuel combustion is the leading environmental threat to global pediatric health and equity: solutions exist. Int J Environ Res Public Health 15(1):16
Ploetz R, Rusdianasari R, Eviliana E (2016) Renewable energy: advantages and disadvantages. In: Proceeding forum in research, science, and technology (FIRST) 2016. Politeknik Negeri Sriwijaya
Satterthwaite D et al (2020) Building resilience to climate change in informal settlements. One Earth 2(2):143–156
Suboticki I, Heidenreich S, Ryghaug M, Skjølsvold TM (2023) Fostering justice through engagement: a literature review of public engagement in energy transitions. Energy Res Soc Sci 99:103053
Sun Z, Zhang F, Yifei W, Shao Z (2023) Literature review and analysis of the social impact of a just energy transition. Fron Sustain Food Sys 7:1119877
Taylor C (1992) Concepts of undervoltage load shedding for voltage stability. IEEE Trans Power Deliv 7(2):480–488
Turok I (2015) Myths and realities of informal settlements: poverty traps or ladders? HSRC Rev 13(3):16–18
Ugarte S, van der Ree B, Voogt M, Eichhammer W, Ordoñez JA, Reuter M, Schlomann B, Lloret Gallego P, Villafafila Robles R (2016) Energy efficiency for low-income households. European Union
UN-Habitat (2015) Informal settlements: habitat III issue paper 22. United Nations, s.l.
United Nations (2023) What is renewable energy? [Online] Available at: https://www.un.org/en/climatechange/what-is-renewable-energy (Access 23 Aug 2023)
Wernecke B, Langerman KE, Howard AI, Wright CY (2024) Fuel switching and energy stacking in low-income households in South Africa: a review with recommendations for household air pollution exposure research. Energy Res Soc Sci 109:103415
Wiese M, van der Westhuizen LM (2024) Impact of planned power outages (load shedding) on consumers in developing countries: evidence from South Africa. Energy Policy 187:114033
World Bank (1999) Non-payment in the electricity sector in Eastern Europe and the former Soviet Union. World Bank, Washington, D.C.
Ziblim A, Sumeghy MG, Cartwright A (2013) The dynamics of informal settlements upgrading in South Africa. Habitat Int 37:316–334

Conclusion

11

Verna Nel and Abraham R. Matamanda

11.1 Introduction

Good urban health and resilient, sustainable and inclusive cities are inextricably linked and are dependent on achieving other sustainable development goals (Giles-Corti et al. 2020). The chapters in this book have also demonstrated the intimate links between sustainable development goals. It is no longer adequate to consider health as a state of disease in the person as it is now acknowledged that health depends on the wider context of life (Corburn 2009, 2013). Good health depends on sufficient income to supply enough nutritious food, plus the knowledge regarding which foods are nutritious (McLoughlin et al. 2020). The lack of clean water and sanitation has many consequences for health and exposes serious gender inequalities (Neumann et al. 2014; Matamanda et al. 2024). Gender inequalities may affect girls' education and have generational repercussions affecting awareness of hygiene and nutrition (Matamanda et al. 2024; Sweetman and Medland 2017). Clean and reliable energy reduces exposure to harmful substances and can limit accidents that burn children or destroy homes, with the trauma associated with the loss of life and property (Kimemia et al. 2014; Kimemia and Van Niekerk 2017; Jaiswal et al. 2022). Decent work enables decent incomes and hence decent food, homes and environments. Good infrastructure—physical and social—supports the economy while innovation stimulates it and job creation, providing decent incomes, and reducing inequalities. Responsible production and consumption, climate action, and regenerative development of terrestrial, riverine, lacustrine and marine environments have many benefits such as better living conditions, sustainable ecosystem services, the limitation of anthropogenic disasters, and the continued ability to produce adequate food (Jagarnath et al. 2020; Shackleton et al. 2018). These all have implications for urban health. Critical to creating and maintaining healthy, resilient and sustainable cities and their environments is good governance with strong institutions.

11.2 Emerging Issues

Chapter 2 reveals the links between poverty, hunger, clean water and sanitation, and health. These links are reiterated in Chap. 3 (see Fig. 3.1). Poverty and hunger are contributors to poor

V. Nel (✉)
Department of Urban & Regional Planning, University of the Free State, Bloemfontein, South Africa

Center for Development Support, University of the Free State, Bloemfontein, South Africa

A. R. Matamanda
Department of Urban & Regional Planning, University of the Free State, Bloemfontein, South Africa

A. R. Matamanda, V. Nel (eds.), *Sustainable Development Goals and Urban Health*, Sustainable Development Goals Series, https://doi.org/10.1007/978-3-031-68734-1_11

health, as poverty affects the quantity and quality (safety and nutritional value) of food consumed. Poor nutrition places can reduce the body's resilience to diseases and increase the risk of cardiovascular disease and diabetes. Children are particularly vulnerable to the effects of poor nutrition as it can stunt physical and mental development. Thus the cumulative effects of poverty and hunger can limit the potential of children raised in environments of deprivation, which in turn, can lead to lasting inequalities, defeating SDG 10 (which envisages reduced inequalities).

Poverty also determines the quality of housing that people can afford (Chirisa and Matamanda 2016; Scheba and Turok 2020). Households without incomes other than social grants have few choices other than backyard rentals, slums or informal settlements with limited access to clean water or sanitation (Mahlakanya and Willemse 2022; Mphambukeli 2015). The lack of decent services in such areas places people at risk of contagious diseases and poor health due to inadequate light, air, space and insulation.

Safe drinking water and good sanitation are critical for good health and well-being and for creating sustainable cities. The lack of these basic services—as is prevalent in informal settlements places residents at risk of waterborne diseases. However, as explained in Chap. 4, women and children are significantly affected by the lack of services, as the responsibility of fetching water usually falls on them, and the increased danger of assault in the evening if they need to go out to relieve themselves. Furthermore, they have additional needs and risks that males do not experience (Matamanda et al. 2024). Consequently, safe water and sanitation have implications for gender equality (SDG 6, and inequalities in general: SDG 10). The realities in the MMM highlight the need for scaling up the efforts and initiatives aimed at addressing water and sanitation issues which greatly disadvantage girls and women tasked with household chores of water provision within the homes. Yet, greater empowerment of women generally leads to enhanced gender equity, with implications for improved prosperity that may have generational benefits (Saleth et al. 2003).

Fires are a constant threat to informal settlements due to the dependence on candles for lighting or paraffin stoves (Pharoah 2012). Besides shack fires, open flames can cause serious burns, particularly to children and pollute the air. These tragedies can best be averted by using safe and clean energy (Kimemia et al. 2014; Kimemia and Van Niekerk 2017). As Chap. 10 points out, fossil fuels can have serious health consequences, causing both pulmonary and other diseases. However, poor households in informal settlements do not have the resources to adopt cleaner energy, nor can they obtain grants or loans as they do not have tenure of the property. This is another means of entrenching inequalities–contrary to SDG 10.

Interrupted electricity supply (load-shedding) has had severe consequences for the economy—ranging from large to micro businesses—and households. Loadshedding combined with the effects of COVID-19 increased the unemployment rate from 35,9% to 40,8% (The World Bank 2024). Many households and small enterprises have had to dispose of spoilt food when refrigerators have failed. Reliable clean energy is also necessary for a good education (SDG 4), especially with the growing dependence on computers or tablets for studying. Furthermore, load-shedding has detrimentally affected the healthcare system (Laher et al. 2019).

Chapter 5 adopts a broad definition of infrastructure that includes social infrastructure such as health and education facilities and economic infrastructure which includes transportation (roads and rail, seaports and airports), energy generation and transmission and telecommunications. At a national level in South Africa, the electrical generation system has been unable to keep up with demand, while there have been crises at major seaports, due to inadequate infrastructure all of which negatively affect economic growth and employment generation. Poor governance is considered to be at the heart of these infrastructural woes (The World Bank 2024).

At an urban level, poor urban streets make walking and cycling unsafe, while inadequate public transport has its hazards (Pirie 2011, 2014). The Auditor General has repeatedly reported poor maintenance and development of infrastructure. Where water infrastructure is not managed, up to 50% of water can be lost, while contamination of water sources through poor wastewater treatment is affecting the health of communities and livestock dependent on the water (Auditor General of South Africa 2022). These pervasive problems have human, animal and environmental impacts and can affect food production, implicating several SDGs. The state of infrastructure, particularly in the informal settlements in the Mangaung Metro is discussed in Chap. 5, revealing deep inequalities in a range of municipal services, namely water, sanitation, waste management and roads. The multiple health issues arising due to the deplorable state of the municipal infrastructure, range from communicable and infectious diseases to the inaccessibility of the settlements for emergency services. However, there are community innovations that demonstrate agency and initiative. Such collaborative actions can lead to improved environments as discussed in Chap. 6.

Unserviced urban settlements create poor environments that impact urban health in many ways—physical, social, emotional and psychological (Rydin et al. 2012), diametrically opposite the cities imagined in SDG 11. Chapter 6 considers the current and potential liveability of previous hostels in Mangaung (see Table 6.5) and the potential for good governance that collaborates with the community to develop a sustainable, resilient safe and inclusive area (SDG 18). Urban liveability encompasses most SDGs related to physical infrastructure and the environment and implies those that affect quality of life (SDGs 1–5, 10 and 16).

Informal housing is particularly vulnerable to climate change (Ziervogel 2021). Most are built from sheets of corrugated iron (colloquially known as 'zinc'), plastic, cardboard and wood (or a combination of these) without sturdy foundations. Due to their relatively flimsy structure, they are easily blown or washed away by storms or floods. They have poor insulation exposing them to summer heat—exacerbated by the health island effect—and the bitter cold of a Free State winter. In many cases they are located on steep slopes or other marginal land (such as low-lying areas prone to flooding) and degraded or unhealthy sites, exposing them to disasters and disease. Furthermore, they generally have poor access to clean water and safe sanitation, as well as to health and educational facilities. Such locations place them at a disadvantage, increasing the inequalities and possible humiliation experienced from living in an informal settlement. This increases anxiety leading to psychological and mental health challenges among the residents of these marginal settlements.

Chapter 7 examines the interaction between housing, health and climate change in informal settlements noting the relationship between SDG 13 (climate action) SDG 3 (good health) and SDG 11 (sustainable cities). It highlights the extreme exposure of such areas to loss of life, property and livelihoods. The chapter notes that while South Africa has excellent policies regarding climate change, these are not being effectively implemented due to a lack of human and financial resources. This problem reflects the need for effective governance to implement policies that take the most vulnerable communities into account.

Well-maintained urban green spaces can improve the quality and liveability of the area (Nieuwenhuijsen 2021; Reyes-Riveros et al. 2021). However, they need to be functional spaces, not 'spaces left over after planning'. Urban green spaces are critical to deal with climate change by reducing temperature extremes and absorbing runoff to reduce possible flood peaks. On the contrary, more hard surfaces increase the amount of run-off and amplify flood peaks. Thus they refer to SDGs 9, 11 and 15. Chapter 8 confirms this relationship. It also points out that urban green spaces in Mangaung are diminishing, despite the need for more such areas, particularly where densities are high. Furthermore, fewer places have access to them.

While unkempt spaces are often viewed as sites of crime and assault and may be occupied or invaded by homeless people, spaces that are adopted by communities offer much potential for their co-evolution with the community to meet its specific needs (Landman 2018). Increasing the number of open spaces and their accessibility can contribute to better ecosystem services, create corridors to enhance biodiversity and manage stormwater more effectively. However, the surrounding communities' attitude and value of the spaces will determine the level of use. The governance of the space through maintenance and co-production with the communities will be key to developing safe and functional greenspaces.

Chapter 9 considers the issues inherent in SDG 16 focussing on the need to sustain peace, justice and inclusive institutions. SDG 16 recognises the role of institutions in advancing socio-economic development, especially through promoting good governance, and accountability while undertaking their service delivery mandates. The chapter highlights the significance of inclusive institutions in ensuring urban health is prioritised through the provision of health infrastructure, basic services and effective use of public funds for community development (see Auditor General of South Africa 2022).

11.3 Concluding Remarks

This book on the SDG and urban health in MMM tackles significant issues relating to the nexus of the SDGs and urban health at a local level. While many studies have focused on the national level and often generalise the realities of the progress and milestones made, this book adopts a local perspective through which lived experiences are explored to gain more insights into the lives of the poor. This approach is especially critical for South Africa considering the high inequalities that at times result in distorted findings, and hence, biased deductions. Instead, the emphasis has been on the realities experienced by the poor urbanites and other marginal groups in urban areas.

The main argument presented in the book focused on the intertwinedness of the SDGs, which called for the adoption of local experiences to gain nuanced insights into the implications of selected SDGs on urban health. The emerging dynamics from MMM show the complexity of some issues, that are best explored from the local level. Hence, there is a need for provincial and national governments to invest in understanding local dynamics.

The issues presented in the book highlight the diversity of the SDGs and the issues emerging from the interrelationships with other areas of development. This supports the Health in All Policies approach that has been applied to understand the dynamics of the SDGs and urban health. Almost all the SDGs have a component of urban health that is uniquely expressed in informal settlements. The interrelatedness of health matters calls for collaborative and participatory action in addressing urban health challenges as well as a concerted effort from different stakeholders. Stakeholders—the public and private sectors, NGOs and community-based organisations—must avoid a silo approach which has and will continue to compromise achieving the envisaged targets by 2030. Working in silos tends to waste resources, and replicate functions while creating engagement fatigue in communities.

While we have attempted to package the SDGs and urban health nexus in MMM in this book, we realise some shortcomings with our approach which only documents the insights from a unique municipality. Hence, we recognise that this book only covers part of the issues albeit contributing to an ongoing discussion considering the SDG targets set for 2030. Hence, we pose some questions and areas for further research and policy action aimed at improving the status quo and increasing the liveability of MMM's low-income areas through a reduction in the environmental, social and economic circumstances in these marginal spaces that compromise people's health and well-being.

- A central urban health challenge in MMM is the issue of urban health governance. We ask whether the MMM has the necessary governance capability and capacity to address the urban health challenges and steer the metro towards addressing the set SDGs within the set time frames. Achieving the SDGs demands robust systems that are inclusive and recognise the voice of the people.
- How should the different issues inherent in the SDGs relating to urban health in MMM and beyond be coordinated? This requires reconsidering urban governance models to those that promote inclusivity, accountability and value human rights.

- Considering the complexity of the SDGs and the thematic issues addressed, it is critical to ascertain how the policy, institutional and spatial frameworks can be employed to guide the achievement of the SDG while reducing urban health challenges.

References

Auditor General of South Africa (2022) Consolidated general report on the audit outcomes of local government MFMA 2020–2021. Auditor General. https://www.agsa.co.za/Portals/0/Reports/MFMA/2020-21/FINAL_MFMA%202020-21%20GR_15%20June_2022%20tabling.pdf?ver=2022-06-15-095648-557

Chirisa I, Matamanda AR (2016) Addressing urban poverty in Africa in the post-2015 period. Perspectives for adequate and sustainable housing. J Settl Spat Plann 7(1)

Corburn J (2009) Toward the healthy city. MIT Press, Cambridge, MA

Corburn J (2013) Healthy city planning. Routledge, London

Giles-Corti B, Lowe M, Arundel J (2020) Achieving the SDGs: evaluating indicators to be used to benchmark and monitor progress towards creating healthy and sustainable cities. Health Policy 124(6):581–590

Jagarnath M, Thambiran T, Gebreslasie M (2020) Heat stress risk and vulnerability under climate change in Durban metropolitan, South Africa—identifying urban planning priorities for adaptation. Clim Chang 163(2):807–829

Jaiswal KK, Chowdhury CR, Yadav D, Verma R, Dutta S, Jaiswal KS, Karuppasamy KSK (2022) Renewable and sustainable clean energy development and impact on social, economic, and environmental health. Energy Nexus 7:100118

Kimemia DK, Van Niekerk A (2017) Energy poverty, shack fires and childhood burns. S Afr Med J 107(4):289–291

Kimemia D, Vermaak C, Pachauri S, Rhodes B (2014) Burns, scalds and poisonings from household energy use in South Africa: are the energy poor at greater risk? Energy Sustain Dev 18:1–8

Laher AE, van Aardt BJ, Craythorne AD, van Welie M, Malinga DM, Madi S (2019) 'Getting out of the dark': implications of load shedding on healthcare in South Africa and strategies to enhance preparedness. SAMJ 109(12):899–901

Landman K (2018) Evolving public space in South Africa: towards regenerative space in the post-apartheid city. Routledge

Mahlakanya I, Willemse L (2022) A living conditions index for main and backyard shacks and backyard rooms in Gauteng's metropolitan municipalities: 2001 to 2011. Town Reg Plann 80:21–39

Matamanda AR, Mphambukeli TN, Chirisa I (2024) Exploring water-gender-health nexus in human settlements in Hopley, Harare. Cities Health 8(1):44–58

McLoughlin GM, McCarthy JA, McGuirt JT, Singleton CR, Dunn CG, Gadhoke P (2020) Addressing food insecurity through a health equity lens: a case study of large urban school districts during the COVID-19 pandemic. J Urban Health 97:759–775

Mphambukeli TN (2015) Exploring the strategies employed by the greater Grasland community, Mangaung in accessing basic services. PhD thesis, University of the Free State

Neumann LE, Moglia M, Cook S, Nguyen MN, Sharma AK, Nguyen TH, Nguyen BV (2014) Water use, sanitation and health in a fragmented urban water system: case study and household survey. Urban Water J 11(3):198–210

Nieuwenhuijsen MJ (2021) New urban models for more sustainable, liveable and healthier cities post covid19; reducing air pollution, noise and heat Island effects and increasing green space and physical activity. Environ Int 157:106850

Pharoah R (2012) Fire risk in informal settlements in Cape Town, South Africa. In: Disaster risk reduction. Routledge, pp 121–142

Pirie G (2011) Sustainable urban mobility in 'Anglophone' Sub-Saharan Africa. Thematic study prepared for global report on human settlements 2013

Pirie G (2014) Transport pressures in urban Africa: practices, policies, perspectives. Afr Urban Rev 133

Reyes-Riveros R, Altamirano A, De La Barrera F, Rozas-Vásquez D, Vieli L, Meli P (2021) Linking public urban green spaces and human well-being: a systematic review. Urban For Urban Green 61:127105

Rydin Y, Bleahu A, Davies M, Dávila JD, Friel S, De Grandis G, Groce N, Hallal PC, Hamilton I, Howden-Chapman P, Lai KM, Lim CJ, Martins J, Osrin D, Ridley I, Scott I, Taylor M, Wilkinson P, Wilson J (2012) Shaping cities for health: complexity and the planning of urban environments in the 21st century. Lancet 379(9831):2079–2108

Saleth RM, Samad M, Molden D, Hussain I (2003) Water, poverty and gender: an overview of issues and policies. Water Policy 5(5–6):385–398

Scheba A, Turok I (2020) Informal rental housing in the south: dynamic but neglected. Environ Urban 32(1):109–132

Shackleton CM, Blair A, De Lacy P, Kaoma H, Mugwagwa N, Dalu MT, Walton W (2018) How important is green infrastructure in small and medium-sized towns? Lessons from South Africa. Landsc Urban Plan 180:273–281

Sweetman C, Medland L (2017) Introduction: gender and water, sanitation and hygiene. Gend Dev 25(2):153–166

The World Bank (2024) Country overview. World Bank. https://www.worldbank.org/en/country/southafrica/overview

Ziervogel G (2021) Climate urbanism through the lens of informal settlements. Urban Geogr 42(6):733–737

The manufacturer's authorised representative in the EU is Springer Nature Customer Service Centre GmbH, Europaplatz 3, 69115 Heidelberg, Germany. If you have any concerns regarding our products, please contact ProductSafety@springernature.com

Printed and bound by CPI Group (UK) Ltd, Croydon, CR0 4YY
01/10/2026
02225987-0016